Fundamentals of the General Theory of the Universe

Vladimir (Waldemar) Groo (Groh)

AuthorHouse™ UK Ltd.
1663 Liberty Drive
Bloomington, IN 47403 USA
www.authorhouse.co.uk
Phone: 0800.197.4150

Fundamentals of the General Theory of the Universe

Possibly for the first time in the history of modern physical science, this book will attempt to step outside the bounds of Einstein's theory of general relativity and quantum mechanics, which were so much a part of scientific thinking in the last century. The author aims to develop a new concept of theoretical physics which is based entirely upon freshly accumulated experimental and observational data. It does not coincide with current theoretical physics. The author believes that this new theory achieves an unexpected but naturally determined and definite breakthrough in natural science.

Published by AuthorHouse 08/27/2014

ISBN: 978-1-4969-8444-9 (sc)
ISBN: 978-1-4969-8445-6 (hc)
ISBN: 978-1-4969-8446-3 (e)

Contents

Introduction

It is generally accepted that history develops in spirals. This is also true for the lengthy and comprehensive development of scientific knowledge. This process can be conditionally classified into several stages.[1]

The first stage covers the period from the origin of natural philosophy until the fifteenth century. During this period, scientific knowledge was undifferentiated, and the world appeared as a single whole. Philosophy was the Queen of the Sciences, and the primary methods of natural philosophy were based upon observations and assumptions. The fields of mathematics, physics, chemistry, and other sciences gradually developed as separate disciplines out of philosophy, and by the fifteenth century they had begun to constitute the separate disciplines of science.

The second stage runs from the fifteenth to the eighteenth centuries. It is characterised by advances in analysis, with further attempts to identify the constituent parts that make up the world and to study them. This stage gave birth to a new scientific method: experimental science. Knowledge was to be acquired through empirical testing. However, the focus was on objects rather than events, and therefore nature was perceived as unchanging and lacking in any dynamics.

The third stage comprises the nineteenth and twentieth centuries, which have been marked by the rapid growth of scientific knowledge and progress. This period is generally characterised as one of scientific synthesis; synthesis was its major methodological principle. More scientific knowledge was gained during this stage than during the entire previous history of science.

At the end of the twentieth century, science stepped into a newly integrated and differentiated stage, which combined synthesis with experiment.

The scientific world view also developed in several stages. At the beginning, the world was predominantly understood as mechanical. This view maintained that if the world has physical laws, then they can be applied to any object or event in the world. This was the world of Laplace, for example.

The electromagnetic world view succeeded this one. It disregarded the notion of a micro world, and was based on the notion of an electromagnetic field and the forces of magnetism, electricity and gravity as studied and discovered by such figures as Maxwell and Faraday.

The quantum world view then stepped in. It focused upon the world's smallest components, the micro world of particles with a velocity equal to the speed of light, and the enormous objects of outer space, a mega world of vast distances and huge masses. The theory of relativity governs this view: it is the world of Einstein, Heisenberg, and Bohr.

The end of the twentieth century saw the appearance of the modern scientific world view. Its basis is that of information, synergy, self-organizing systems (for both organic and inorganic nature) and probability theory. This is the world of Stephen Hawking and Bill Gates – a world of space wrinkles and artificial intelligence.

1

A Past Scientific World Best Forgotten?

1.1 Practical experience as the criterion of truth

Scientific knowledge seldom develops in a clearly defined path. Its course is often marked by sharp turns, zigzags, and even dead ends. Quite often, fundamental notions are questioned and then completely redefined.

Granted that practical experience provides one of the criteria for truth, in the course of time the close observation of phenomena or objects yields an accumulation of empirical evidence. Patterns are discerned, and new features are discovered, which ultimately lead to completely new ways of thinking about established notions. Sometimes changes are introduced into a theory used to describe a phenomenon or object. Although such examples are relatively few, one example from the scientific literature should be enough to illustrate such a change.[2]

When his general theory of relativity was completed, Einstein applied it to the universe. His interest was not in such specifics as the positions of certain stars or planets. He believed that his theory and the application of its equations would describe the structure of the universe as a single whole.

As little was known about the distribution of matter in the universe, Einstein had to make certain assumptions. He made the simple, working assumption that matter is distributed homogeneously throughout the universe. Of course, there are local variations in homogeneity, which differs where the concentration of stars is either higher or lower than the average. However, given the scale of the universe, it can be considered completely homogeneous for the purposes of Einstein's assumption. Further, the assumption implies that all places in the universe are pretty much the same: the Earth, for example, does not occupy a special place within the cosmos. Einstein also assumed that the universe is isotropic – that is, that it looks practically the same from any location.

Finally, Einstein suggested that, on average, the properties of the universe are constant. In other words, the universe as a whole is static. Although Einstein had few observations to verify and support his hypothesis, the idea of an eternal, stationary universe was a theoretical postulate for his position.

Having arrived at his sought-for model of the universe, Einstein went on to try to find solutions to his equations – solutions that would, if found, describe a universe with the characteristics that his theory had predicted for it. However, it did not take him long to discover that his theory did not provide for such solutions. The reason was very simple: the mass distributed throughout the universe was *not* uniformly distributed. On the contrary, one mass tended to fall onto another due to their mutual gravitational attraction.

Einstein was both puzzled and perplexed by this result. After a further year of hard work, he decided that the equations of the general theory of relativity could be modified to allow for the assumption of the existence of a static universe.

He found that it was possible to include an additional term to his equations without nullifying the physical foundations of his theory. The additional term provided for a spatial void – that is, a vacuum, which would have a non-zero tension of energy. It follows that every cubic centimetre of empty space has an amount of energy and hence also has mass. Einstein called this constant energy-density of the vacuum in the universe *the cosmological constant.*

The mathematical structure of Einstein's equations determines that the tension of the vacuum exactly equals its density of energy and, therefore, that the tension is defined by the same cosmological constant.

The energy and tension of the vacuum cannot be detected, excluding the case of the gravitational effect.[3]

The gravitational effect of the vacuum came as quite a surprise. According to the general theory of relativity, pressure and tension contribute to the gravitational force of massive bodies. When you squeeze an item, its gravitational attraction increases; when you stretch an item, its gravitational attraction decreases. Usually this effect is insignificant. However, if you manage to continue stretching the item without it disintegrating, it is possible to decrease its gravitational attraction up to a point of complete neutralization or even repulsion. It is exactly the same with a vacuum. The repulsive gravitation of the tension of a vacuum by far exceeds the gravitational attraction of its mass. This eventually leads to gravitational repulsion.

Einstein needed this conclusion to solve his problem. With it, he could choose any value of the cosmological constant to ensure equality between the gravitational attraction of matter and the repulsive gravitation of the vacuum, with a static universe as the outcome – in keeping with his predictive model. Using his equations, Einstein deduced that a balance is achieved when the cosmological constant is equal to half the density of the energy of the substance.[4]

The revised equations lead to a striking consequence: they imply a static universe with space that is curved and closed in on itself and that has the overall surface, volume, and shape of a sphere. A spacecraft travelling straight ahead in such a universe would finally return to its starting point. This closed space is a three-dimensional sphere. Its volume is finite, although it has no boundaries.

Einstein described his closed model of the universe in an article in 1917. [5]He admitted that he could not confirm by empirical observations the non-zero value of the cosmological constant. Indeed, the only reason that he introduced the constant was to justify the idea of a static universe. Ten years later, after evidence emerged that the universe was expanding, Einstein regretted his "cosmological constant" hypothesis, referring to it as the biggest blunder in his career.[6] After this unsuccessful debut the notion of repulsive gravitation disappeared for almost half a century from the physical studies, only to re-emerge much strengthened.

Einstein, to his credit, admitted that his scientific idea was erroneous, an admission confirmed by the provisions described here as well as by practical experience. Despite the time lost to scientific mistakes, it is pleasing that there is now room for dissenting opinions.

The situation is different for another of Einstein's claims whose application has caused difficulties for its author ever since he first introduced it.

It is the claim that denies the existence of ether.

1.2 Einstein and the theory of ether

In a scientific article in 1905, Einstein reviewed two principle postulates: the general principle of relativity and the invariance of the speed of light.[7] The adoption of these postulates immediately involved the Lorentz transformations (not only for electrodynamics), the contraction of length, and the relativity of simultaneity. In the same article, Einstein pointed out that, because no physical attributes could be assigned to ether, it was not necessary to posit it. He also noted that other dynamic properties were taken into account as part of the kinematics of the special theory of relativity. Ever since, the electromagnetic field has been viewed as an independent physical object, and not as an energy process in ether.

The introduction of this notion caused a number of difficulties. Einstein had experienced some problems (as shown in the foregoing example) when his findings showed that he needed to add a certain amount of energy (hence mass) to the empty space of the universe.

Other reasons for denying the relevance of the concept of ether included its ambiguous attributes: the impalpable character of the supposed substance; its transverse elasticity; its extraordinary wave-propagation velocity when compared with gas, liquid, etc. Also, there was the difficulty of proving the quantum nature of the electromagnetic field, which is incompatible with the "continuous ether" hypothesis.

Later, after his creation of the general theory of relativity, Einstein revived his use of the idea of ether by stipulating his own meaning for the notion. According to his new meaning, "ether" refers to "physical space" as defined by the general theory of relativity. Unlike luminiferous ether, Einstein's physical space is not substantial. For example, it is impossible to assign self-motion and an identity to points in such a space. This notion of space, unlike the notions used by Lorentz and Poincaré, is in keeping with the general theory of relativity. However, most physicists have chosen not to revive the obsolete term.

1.3 An overview of the concept of ether

According to a few works of the earliest Greek thinkers, "ether" meant "upper air" ("hypothetic all-permeating medium" in Latin). Ether was understood as a special celestial object in the cosmos – that is, as a space-filling substance.[8] Plato, in his dialogue *Timaeus*, writes that God created the world from ether, while Lucretius Carus, in his poem "*On the Nature of Things*", mentions that "Ether feeds the stars". That is, that celestial bodies consist of condensed ether.

Aristotle describes ether in more detail. He believes that planets and other celestial bodies consist of ether, which is the fifth element of nature – the quintessence. Moreover, unlike the other elements (earth, air, fire and water), ether is constant and permanent. For Aristotle, the Sun is not made of fire, but is a huge cluster of ether. The sun radiates heat by influencing its ether as it revolves around the Earth. Ether also fills the region of the universe above the terrestrial sphere, staring from the Moon's sphere. So, according to Aristotle, ether transfers light and heat from the Sun, and light from the stars.

Medieval scholastics adopted Aristotle's understanding of ether, an understanding that remained unchanged until the seventeenth century.

In 1618 René Descartes put forward a hypothesis of luminiferous ether, which he further developed in his *Principles of Philosophy* (1644).

According to his natural philosophy, Descartes regards ether as a fine matter, similar to liquid, which determines the patterns of light distribution. For Descartes, ether fills the entire matter-free space of the universe, but it does not impede the movement of physical objects through it. Descartes, like Aristotle, does not recognize empty space as such.

As with any celestial matter, Cartesian ether is in a constant state of motion, primarily a swirling motion. Centrifugal force and an emerging, mutual pressure of particles throw ball-like ether particles off the source. An observer perceives this motion as light distribution. Descartes believed the speed of light to be infinite. He also developed his own theory of colour. According to his theory, the different speeds of rotation of ether particles generate the different colours.

In 1690 Huygens published his *Treatise on Light,* in which he expanded considerably on Descartes's teaching on the subject. Huygens hypothesised that light travelled in waves in ether, and he developed his mathematical formulae for wave optics.

The following unusual optical phenomena were discovered at the end of the seventeenth century and accommodated within the luminiferous ether model:

- diffraction (by Grimaldi, 1665);
- interference (by Hooke, 1655);
- the double refraction of light (Bartholin, 1669 and later studied by Huygens);
- the determination of the speed of light (Romer,1675).

As such findings began to accumulate, two dominant and contrasting models of the nature of light began to emerge.

1) The *emission,* or corpuscular, theory of light proposes that light is made up of small particles emitted by a source. This hypothesis was supported by the evidence that light travels in a straight line, which is the physical presupposition of geometrical optics. However, diffraction and interference did not fit with this theory.

2) The *wave* theory of light suggests that light travels in waves as a result of outbursts of ether. It is worth noting that in those times a wave of light was thought of as a single pulse, not the infinite periodic

vibrations of today's scientific theories. For this reason it was very hard to explain light phenomena using the wave theory.

The Descartes-Huygens concept of luminiferous ether gradually came to be generally accepted in scientific circles. The seventeenth and eighteenth centuries' disputes between supporters of the Cartesian and atomic theories, and those who contested the merits of the emission and wave theories, did nothing to compromise the concept of luminiferous ether. Even Isaac Newton, who favoured the emission theory, conceded that ether was involved in light effects.

By virtue of Newton's scientific achievements and authority, the emission theory of light became the dominant theory in the eighteenth century. Ether was regarded as a bearer of light particles. Light refraction and diffraction were explained by the changes in the density of ether occurring when light impinged on bodies (diffraction) or passed from one medium into another (refraction). In the course of the eighteenth century, the notion of ether as a constituent of the universe gradually receded into the background of scientific thinking. However, the theory of ether vortices lasted and was unsuccessfully employed to explain both magnetism and gravitation.

At the beginning of the nineteenth century, the wave theory of light, which defined light in terms of waves in ether, produced a decisive victory over the emission theory. As early as 1800, the emission theory was attacked by Thomas Young, an English scientist. He was the first to propose and coin the expression "a wave theory of interference" based on his own principle of superposition. On the basis of his experimental findings, Young provided accurate assessments of light wavelengths for various spectra of coloured light.

At about the same time, Augustin-Jean Fresnel performed a series of ingenious experiments demonstrating that purely wave effects could not be explained by the emission theory.

From the beginning, Young and Fresnel regarded light as elastic, longitudinal vibrations of rarefied but very resilient ether, similar to that of sound in the air. They held that any light source causes elastic vibrations of ether. The frequency of such vibrations is huge and is not recorded anywhere else in nature. The vibrations propagate at tremendous speed, and any physical object draws ether in; the ether penetrates it and concentrates inside it. The index of light refraction depends on the density of ether in a transparent object.

However, the polarization effect still called for an answer. In 1816 Fresnel had suggested that the light vibrations of ether were not longitudinal but transverse. This would easily explain the polarization. Earlier, however, transverse vibrations had been recorded only in incompressible solids, while ether was considered to have properties similar to those of gases or liquids.

The advent of the nineteenth century saw a sharp growth of interest and trust in the concept of ether.

After it became clear that light vibrations are strictly transverse, it was necessary to establish the properties ether would have to have in order to allow for transverse vibrations and exclude longitudinal ones. In 1821, Henri Navier developed general equations to describe the propagation of disturbances in an elastic medium. Navier's theory was expanded by Augustin-Louis Cauchy, who stated in 1828 that, generally speaking, longitudinal waves should also be present.

Fresnel hypothesised that ether was incompressible, but provided for transverse shearing. But such an assumption is hardly reconcilable with the complete permeability of ether to substance. George Gabriel Stokes addressed this problem by likening ether to resin. During quick deformations involving light emission, it behaves like a solid, whereas during slow deformations involving planetary movement, for example, it is plastic.

Fresnel even suggested that ether consists of particles relatively similar in size to the length of a light wave.

After Maxwell presented his equations on classical electrodynamics, the theory of ether was reconsidered. Maxwell neither developed specific models of ether nor took into account any other properties of ether in his calculations, except for ether's supposed ability to maintain the displacement of a current – that is, electromagnetic vibrations in space.

When H. Hertz proved experimentally Maxwell's theory, ether was confirmed within the prevalent scientific outlook as the general medium for light, electricity, and magnetism. When wave optics also became part of Maxwell's theory, there were expectations that it might act as a foundation for the construction of a physical model of ether. Many of the world's most prominent scientists conducted studies, researches, and experiments in this sphere. Although some of them, such as Maxwell himself, Umov, and Helmholtz, mentioned the properties of ether, they were primarily studying directly the properties of the electromagnetic field. However, other prominent scientists – George Gabriel Stokes and W. Thomson – worked on the understanding of the nature and properties of ether itself. They sought to determine its pressure and the density of its mass and energy, and to connect it with the emergent atomic theory. Attempts were also made to relate ether to gravitation, but even Maxwell himself was unable to make any significant progress in this direction.

In one of his hypotheses, D.I. Mendeleev drew attention to the fact that a particular state of strongly rarefied gas might be either ether or some unknown, inert gas of negligible weight. That is, that it might be the lightest chemical element. In 1871 Mendeleev wrote on a sketch of the periodic table in the *Principles of Chemistry*: "Ether is lighter than anything, by millions of times". He added in a notebook in 1874: "Air with zero pressure has some density, which is the ether." However, these ideas were not reflected in his publications at that time. The discovery of inert gases at the end of the nineteenth century took forward attempts to further identify ether as a chemical substance. At William Ramsay's suggestion, Mendeleev added a zero group to the periodic table, leaving a place for elements lighter than hydrogen. He thought that the group of inert gases could be enlarged by adding coronium and also what he took to be the lightest, still unknown element. He called this element "newtonium" and thought of it as making up the domain of ether.

The history of the concept of ether shows that the majority of great scientists assumed its presence while making their discoveries and taking science forward. It is natural that when any body of knowledge develops, obsolete notions that impede its progress are discarded. Except that the abandonment of the concept of ether is best likened to putting aside the wheels of a cart or the runners of a sledge: it is self-defeating.

2

Ether, Energy, and Matter

2.1 A rationale for the structure of ether

Taking into account the history of scientific discoveries and advances and their practical confirmation, it is most likely that the premature denial of the existence of ether is yet another scientific delusion. We, the authors, are certain that ether exists and therefore recommend this book to the attention of those interested in the question of the existence of ether.

To facilitate the reader's comprehension, we suggest a "from simple to complex" method, an approach seemingly both reasonable and feasible. In the first place there needs to be a shared understanding of the basic concepts of matter and energy. This will employ the accumulated knowledge about the laws of interaction of material bodies, mathematical laws, logical ways of thinking, and previous conceptions of the medium of ether. Obviously, speculations about the structure of ether will be based upon data obtained about the interaction and properties of material bodies.

The material world is made up of particles of material bodies detected either directly by the senses or with devices to aid the senses. Ether, unlike the material world, is a multiple-level substance consisting of particles which are not detectable by modern devices.

The discrete structure of material particles and bodies can be observed, which gives grounds for considering that ether as a medium might likewise have a discrete structure, possess viscosity, be unstable in space and time, have changeable densities of its constituent particles, and be located in a medium of finer particles. It is not surprising that there are particles of different levels in different parts of space. To simplify things, let's imagine a glass container filled with stones. The voids between the stones are filled with sand, the voids between the grains of sand are filled with silicate dust, the voids between particles of silicate dust are filled with water in which oxygen is dissolved, et cetera. If ether is likened to this example, the structure of ether has no voids or gaps because there are always particles of different types to fill such voids and gaps. At the same time, if we set the boundaries for the distribution of ether, we must acknowledge the presence of absolutely "nothing" and be required to answer the question, "What is it and what is next?"

For the purposes of explication, let's establish some basic notions and definitions:

1) *The material level* of ether comprises the totality of particles of electrons, protons, neutrons, and atoms of the chemical elements, and of larger material objects. The particles form ball shapes and tend to synthesise or form chemical bonds.

2) *The photon level* of ether consists of photons which, like atoms, have a complex structure, so form ball shapes and also tend to synthesise and form chemical bonds. This is a description of the primary energy process of interactions within the solar system.

3) *The gamma-photon level* of ether is of an even finer level; its particles are much smaller than photons, but keep the same properties. This is, in its turn, the primary energy process of interactions within the Galaxy.

It should always be borne in mind that big things are made up of smaller ones. A good example of this would be the growth of a tree trunk from chemical elements, due to the process of photosynthesis. An opposite process would be the resolution of a part of the trunk (for example, chopped wood) into volatile, gaseous products resulting from the process of burning.

The properties of particles of the finer level determine the physical properties of the objects at this level.

Ether particles can be distributed in space in a variety of ways. Therefore, it is appropriate to consider the structured and non-structured ether of a given level.

Structured ether at the photon level (hereafter "structured ether") consists of photons located in space in such a way as to form a lattice with equal distances between its points. Photons tend to appear as a ball, which is a minimally complex spatial object having maximum rigidity.

Non-structured ether at the photon level (hereafter "non-structured ether") is unstable, and the distances between photons are different.

The photons of structured ether have an energetically favourable spatial structure. However, it is practically impossible for structured ether to reach this state if it is subject to constant, external disturbances, though photons do tend to line up to eliminate the effects of any disturbance.

It was mentioned above that ether is a multiple-level substance. This quality makes it rather difficult, and often impossible, to study all the processes that go on inside it at the same time. The most reasonable path forward is to confine our research to the photon level in order to review some primary interactions in ether.

We believe that the existence of the "photon field" structure can be substantiated by the following considerations:

1) Electromagnetic radiation of a photon field is uniformly propagated, possibly because of the equal distance between the lattice points.
2) The structure of the photon field does not have redundant bonds. It has, therefore, a minimum integer number of points in the spatial lattice.

With regard to the internal structure of a photon, we assume that it consists of a minimum number of elements, namely three. Geometry teaches us that one point does not determine a space, that two points can determine only a plane, and that it takes at least three points to determine a space. Deuterium is the simplest chemical element consisting of three particles.

Presumably the structure of a photon is similar to that of deuterium. Therefore, under certain conditions such as excess pressure, high temperatures, etc., synthesis should be possible. This will be re-visited later. The pressing topic now is, "How does the photon field transfer energy?"

The mechanism for the propagation of electromagnetic disturbances (light) in the photon field is as follows. Electromagnetic disturbances start from the transverse motion of a charge in the photon field structure under an external influence. This motion affects the next charge along the direction of propagation, but a charge with a different sign in accordance with Coulomb's law. Maxwell's displacement currents are also generated, which run along the charges' motion in the same direction, but have different signs. Hence, magnetic intensity occurs between currents in a perpendicular direction. This intensity is the sum of two or more magnetic intensities of displacement currents. Besides creating a magnetic component orthogonal to the electric tension (charge motion), the resultant magnetic field acts like a damper that limits the speed of light distribution. Thus, bond charges of the dipoles re-transmit electromagnetic waves. The light coming to an observer is not an initial phenomenon or an emitted source photon, but the photon that was retransmitted several times.

At the same time it should be noted that the photon field is always ready to quickly transmit a disturbance in any direction and for any distance because of its being in a resonant mode. To use a simple analogy, it could be described as a gelatin dessert, which vibrates in response to any impact.

2.2 The properties of ether

It was stated earlier that ether in general, and the photon field in particular, is a substance distributed in space and that this distribution can be heterogeneous. Bonding between fluctuations of density (viscosity) naturally causes resistance for the motion of material particles in the medium of ether. But most important is the fact that, by analogy with a material medium, the substance in question will possess internal potential pressure in accordance with known physical laws. In a small area of structured ether, such pressure acts equally in all directions. Vectors of force generating internal potential pressure can be activated by a vector of outside force in non-structured ether.

Internal potential pressure as inertia characteristic of medium resists any injection, and this determines the internal potential compression energy. Density changes are possible within ether. Hence, it is natural to assume that particles at a finer level would make for areas of less density. Which raises the question, "How fast does the pressure equalize in such areas?" This index of pressure equalization can be considered one of the fundamental properties of the ether medium because it is connected to its compression elasticity. Internal potential pressure influences many processes of the interconversion of matter and energy. However, let's first of all distinguish the formation of material particles from the medium of ether.

Various scientific researches and ordinary life experiences support the role that resonance can play in affecting material conditions. A few examples involving the breaking of material objects should suffice: the collapse of a bridge caused by a platoon of soldiers marching in step; Shalyapin, the opera singer, cracking an empty glass after singing top notes of a certain acoustic frequency; and, more specifically, the conversion of material particles under laboratory conditions.

However, the conversion and, naturally, the formation of material particles from ether can take place not only under laboratory conditions, but in real life as well. Every day, in real life, we witness this process and take it for granted without attending to details, and yet these instances of practical experience provide the criterion for the truth of this claim.

When we note such information as, for example, that the mass of the Earth is constantly increasing, we understand that this is caused by a transfer of energy. However, the explanation of the cause of this energy transfer seems incomplete if the hypothesis of the creation of matter from the medium of ether under the influence of special catalysts – energy, internal potential pressure, etc. – is not examined.

The distribution of an electromagnetic disturbance (light) in a vacuum is usually taken as an example. However, what is observed seems somewhat different if all the phenomena described as acting within a vacuum are examined not in a vacuum but in the medium of ether. In the case of ether, the emergent magnetic field – which, among other things, acts like a damper limiting the speed of the distribution of light – becomes able to create the conditions for the local accumulation of energy. Such zones of the accumulation of local energy, together with the internal potential pressure, can be exposed to electromagnetic disturbances from several directions – that is, from one or more sources by changing the direction of the disturbance (the refraction of light rays) and possible resonance phenomena. The totality of all these components provides conditions for the creation of electrons, protons, neutrons – that is, material particles. In each such case, non-material photons serve as the construction material.

Taking into account that in real life, areas of a photon field receive light disturbances of different intensity, at different angles, etc., it is obvious that some areas can not only create material particles but that more often they intertwine under the influence of vector disturbance, and thus accumulate more energy. This is very important for the creation of matter, but the lifetime of matter depends upon many conditions. However, for the moment, the focus is upon the process of the creation of matter.

Newton canonized the statement "Ether = matter, ether = energy, ether = God".

It is well known that the representation of an elementary nuclear particle takes the form of a ball. This is worthy of further consideration.

The material world that we observe today is at the stage of the conversion of structured ether into material particles: electrons, protons, and neutrons, which in turn create hydrogen atoms. Given that this discrete process occurs locally in space, it seems obvious to follow that under the force of the compression of structured ether particles there are local compressions of developing hydrogen clusters.

3

Hypotheses about the Manifestation of Gravitation and Forces of Inertia

3.1 The nature of gravitation

Gravitation is a familiar phenomenon of the human condition, and even more so for those cosmonauts who venture into outer space. This is one important reason that studies of gravitation spark a lot of interest.

Newton applied mathematics to formulate physical laws of gravitation and the forces of inertia. Einstein took the next step. He explained gravitation as the form of curvature of the empty space near material objects and inertia as an equivalent of gravitation. According to Newton, acceleration is absolute and determined by ambient space. When thinking speculatively about the absolute or relative character of acceleration, Einstein backed Mach's principle according to which inertia is determined by the totality of masses in the universe. However, this principle is challenged by Mach's own paradox where inertia is concerned: that a rotating object standing without the universe cannot withstand centrifugal forces. This impasse means that it is completely unclear how the mass of the universe creates inertia.

It is generally agreed by many scientists that the empty space-time curvature theory is self-sufficient, and that there is no such problem as the nature of gravitation. However, even from a philosophical point of view, this approach is not convincing. According to current ideas, the physical vacuum of theories plays a special role in all interactions, except that there are instances of gravitation in which gravitons function as exchange-field quanta that have not been fully studied.

If the existence of ether is granted, it makes it easier to answer many of the questions relating to, and to meet many of the challenges of explaining, the nature of gravitation. But the question of from which viewpoint it is reasonable to tackle these challenges is anything but simple.

As mentioned, the medium of ether has internal potential pressure, which acts equally in all directions within a limited area under consideration.

It is also the case, for example, that the mass of the Earth is constantly increasing, and this might be a manifestation of the effect of the properties of ether. In view of the fact that there are no voids and gaps in the medium of ether, either in outer space or near the surface of planets, including the Earth, it is clear that internal potential pressure acts everywhere. The finer level of ether brings pressure upon larger objects. When dealing with this question it should be specially noted that the medium of ether should not be equated with air or with air-free space, the distribution of which depends upon the medium of ether.

When analysing this issue, it is necessary in the first place to classify objects in space with regard to their ability to generate the particles that make up material bodies.

To start at the beginning: the centre of a Galaxy (a black hole) absorbs all kinds of radiation from bodies in space, as well as absorbing all the bodies in space themselves and other material bodies. A black hole also produces and radiates the finest energy-intensive, gamma-photon radiation, which has the highest frequency in the Galaxy.

Different deformations at the gamma-photon level, and its internal pressure, create conditions for the synthesis of gamma-photons.

A star, such as the Sun, absorbs gamma-photon radiation produced by the centre of a Galaxy, and synthesises and produces photons, thus forming the photon level and creating the simplest chemical element: hydrogen.

A planet, such as the Earth, absorbs photons produced by a star, such as the Sun, and synthesises matter. The synthesis of photons for the creation of matter is possible due to the potential pressure and resonance phenomena at the photon level. In turn, they are caused by the gamma-photon level, which acts inside and around the photon level. This last condition is very important for the creation of matter and elementary particles.

Figure 3.1 outlines all the processes related to the absorption, synthesis, and production of elementary particles by objects in space.

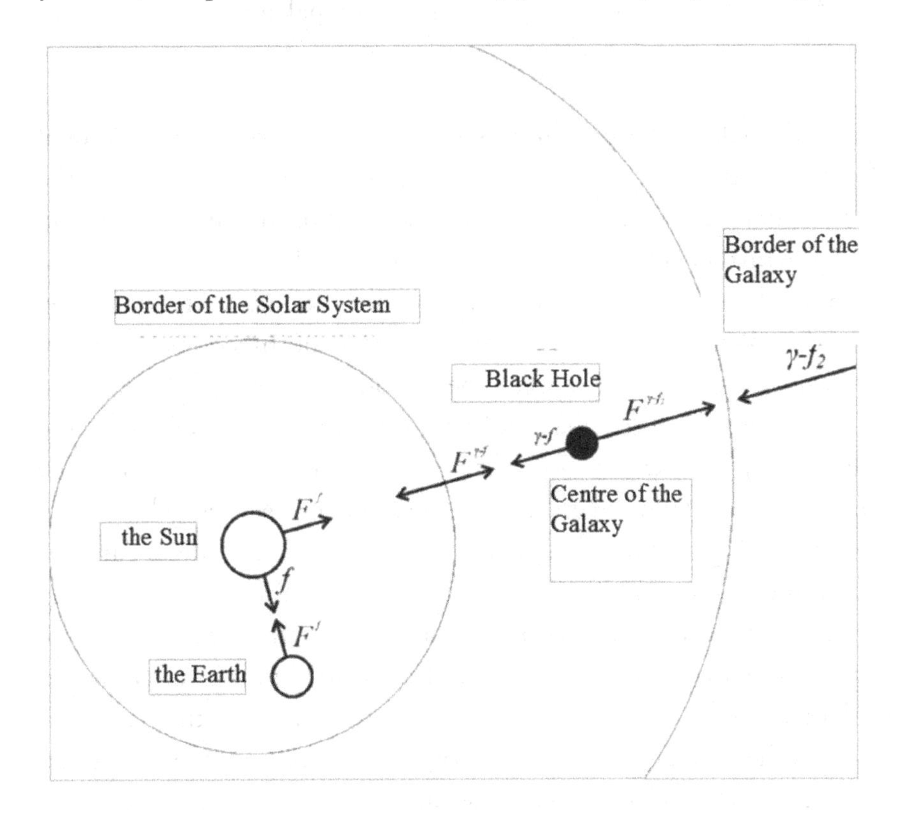

Fig. 3.1. The creation of the attracting components of gravitational forces of space objects.

Where:

γ-f_2 is the radiation of the gamma-photons of the second level travelling to the centre of the Galaxy from the Metagalaxy;

γ-f is the radiation of gamma-photons produced by the centre of the Galaxy;

f is the radiation of photons by the star (the Sun);

$F^{\gamma-f_2}_{grav.}$ is the component of the gravitational force by gamma-photons of the second level;

$F^{\gamma-f}_{grav.}$ is the component of the gravitational force of the gamma-photon level;

$F^{f}_{grav.}$ is the component of the gravitational force of the photon level.

Let's look at the preconditions of the formation of gravitational force. The centre of the Galaxy receives gamma-photon radiation of the second, finer level; absorbs it; and synthesises gamma-photon radiation with the further emission of gamma-photons. Taking into account that gamma-photons are connected by a chemical bond, it is noteworthy that they generate equal potential pressure in all directions. A certain imbalance is quite possible when the "absorbing capacity" of the centre of the Galaxy exceeds the "radiation capacity" of the centre of the Metagalaxy. As there are no voids or gaps in the medium of ether, the centre of the Galaxy will gravitate to the centre of the Metagalaxy using the "attracting wire rope" along the direction of absorption of gamma-photons of the second layer.

It we take the gamma-photon level, we can observe the same situation between the centre of the Galaxy, which emits gamma-photon radiation, and a star, such as the sun, which absorbs this radiation and synthesises it into photons with the further emission of photon radiation. When this happens, the "attracting wire rope" of gamma-photon level is formed.

According to the concepts outlined, the "attracting wire rope" of the photon level can be formed between a star, such as the sun, and a planet, such as the Earth, when the "absorbing capacity" of a planet exceeds the "radiation capacity" of a star.

What follows is the probable formation of such "attracting wire ropes" between space objects that makes for their mutually influenced arrangement.

Together with the "attracting component" of gravitation, there is presumably a "pressing component" of changing, potential ether pressure. This pressure might be caused by the following conditions.

Under excess pressure, photons with a capacity for synthesis ensure that the photon level of ether participates in the creation of matter and an increase of the mass of objects, also under certain conditions. It is also likely that some part of the photon-level ether will flow over to the material space object. In the process of being created, new material builds up and photons condense and, in consequence, some areas are freed. By analogy, this phenomenon is comparable to the effects of a thermonuclear explosion. At the initial stage of the explosion, when charged components go through synthesis, nearby objects – trees, light civil structures, vehicles, etc. – are drawn into ground zero. Note that the greater the mass of a material space object, the greater the quantity of photons absorbed. It is quite possible that a material space object which absorbs photons emits radiation. However, at this stage of our research, we are more interested in the "consumption" of photons.

Thus, if some photons flow from the medium of ether to the material space object, and if there is an automatic redistribution of the photon level along the surface of an object to equalise the medium pressure based on local "consumption", it is impossible to refer only to the internal potential pressure of ether. It is clear that when near the object in space, ether not only acquires internal, potential pressure, but also follows the laws of flow in a medium. Bernoulli's principle can be applied to it given a possible similarity to material bodies. The total speed of the photon flow near the surface of a space object will be a combination of a component perpendicular to the surface caused by an "inflow" of photons from the ether medium, and a component parallel to the surface caused by an overflow of photons due to the uneven "consumption" of photons by various parts of the object.

It is quite possible that the total speed of the flows includes the speed of a twisting rotation. According to Bernoulli's principle, applied by an assumed analogy with material bodies, if the total speed of the flow near the surface of the space object increases, its potential pressure decreases for a given section of the object. The flow of photons from a general mass close to a space object is predetermined by the differential of internal potential pressures at a distance and close to the space object surface in the ether medium. When approaching the material body surface, the speed of the photon's flow will be constantly increasing and based on the intensity of "consumption", which is itself directly connected with the mass of the space object. The increase of the speed of flow of a photon upon approaching the space object is indicative of the force of acceleration imparted by the ether medium. These are all the components comprising the appearance of the "pressing" force. Thus, take a material space object with a certain mass, such as the Earth. In a state of relative rest, ignoring the forces of interplanetary interactions, the planet will be exposed to the influence of a photon field flow of a certain acceleration. The product of the mass of the material space object and the acceleration of the flow of photons results in a certain force value – the "pressing component" of the potential force pressure. The "pressing component" of a force pressure will be in the same direction as the acceleration of the photon's flow, which is directed at the centre of the space object. Any material items located near the space object will be pressed towards this object by the action of the "pressing component" of forces of potential pressure. Their acceleration will be equal to the acceleration of the photons of the ether medium, which depends upon the mass and size of the space object.

For the purpose of the proposed research, it will be assumed that the gravitational force of material-level objects results from a synthesis of matter from particles of the finer photon level. It follows that gravitation can be provisionally defined as a force that appears as a result of the synthesis of particles of gamma-photon and photon levels in the formation of material space objects, and that there is a release of energy from this process at a different frequency in the manner of a slow process that is something like that of a nuclear explosion.

This also means that such a synthesis serves as a condition for the formation of a gravitational field. And that because synthesis goes on at all levels, the so-called gravitational forces are the same as nuclear forces.

3.2 The analysis of inertia

The foregoing analysis of gravitation showed that, allegedly, this phenomenon is caused by the finer gamma-photon and photon levels.

Logic suggests that the explanation of inertia is to be found at the same levels. To perform a thorough and objective study, it is reasonable to consider any specific body with mass m and volume V as a sum of ultimate particles with mass m_i and volume v_i. Note that the sum of masses m_i totals m, and the sum of volumes v_i totals volume V.

$$\sum_{i=1}^{n} m_i = m; \sum_{i=1}^{n} v_i = \mathrm{V} \tag{3.1}$$

We have, then, the same specific body that is virtually divided into ultimate components.

With any attempt to move the body, there will be an initial resistance that occurs at the outset – that is, the resistance will be of a force equal and opposite to the applied force.

When the speed of the moving body becomes constant, the "retarding" force disappears and only appears again when there is an attempt to change the speed of movement through the application of an external force. Thus the "retarding" force appears only if a body's acceleration changes. It can even be visually determined that the properties of air have no influence because the drag due to aerodynamics is negligible at the speed of such motion. The experiment can even be carried out in a tank filled with rarefied air. The result is the same. One might think that it is impossible to analyse this hard-to-explain phenomenon. But it is no coincidence that at the very beginning of the research, the body – the subject of the research – has been virtually assembled from multiple, ultimate components.

Under the effects of acceleration, each virtual component will increase the number of interacting particles and have an effect on the space lattice of the ether photon level, which also has internal potential pressure. The totality of all these active components requires additional energy to overcome resistance, which finally leads to the appearance of a "retarding" force – inertia.

If one tries to apply external force to stop an item that moves at some constant speed, then everything happens in a reversed sequence.

Analysing the reasons for inertia, it becomes clear that, just as with gravitation, inertia is determined by the finer photon level and the mass of the body in question. The volume of a body has no influence on the process because the aerodynamic drag of a body does not manifest itself at the photon level.

4

An Analysis of Electromagnetism

4.1 The properties of electric currents

The classical definition of an electric current implies the ordered movement of electrons in a conductor of an electric field. We believe an electric current is not only a movement of electrons but also an electromagnetic wave in the photon field of the conductor.

An electric current depends upon the properties of the electromagnetic field. Figure 4.1 shows an electric field conductor.

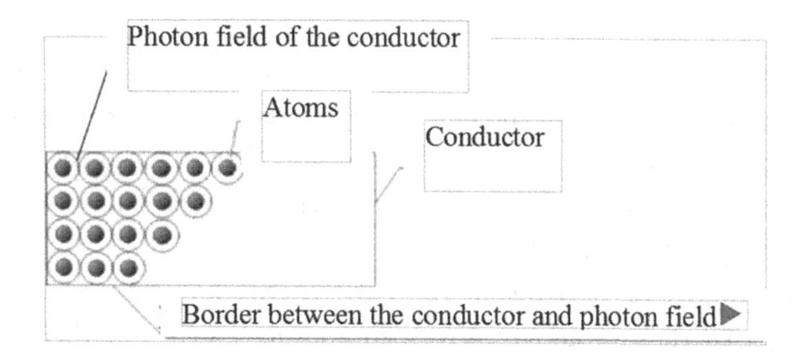

Fig. 4.1. The structure of an electric field conductor.

A conductor consists of atoms connected by electron clouds. Let's consider an atom of the conductor as a particle with properties different from the surrounding medium, the photon field of the conductor. The number of electrons at the outer energy level determines the electrical properties of the conductor.

As the release of energy is discrete, let's regard the movement of an electric wave in the photon field as a non-linear function (oscillations in a medium). When electron orbits tend to their initial state, forces appear that are perpendicular to the forces exciting the electric current but with a minus sign (an anti-phase) (see Fig. 4.2).

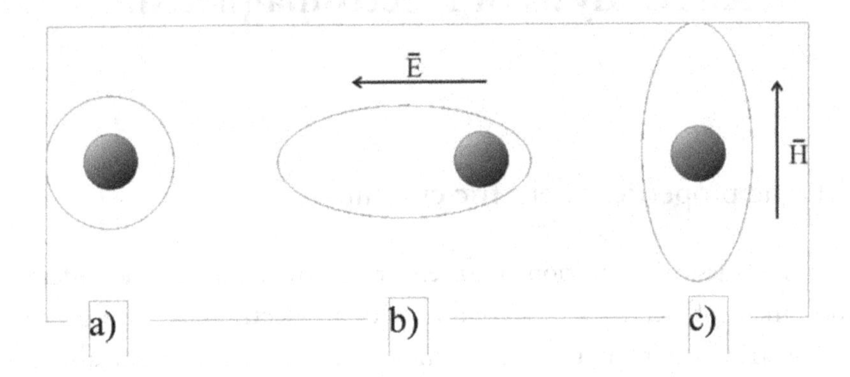

Fig. 4.2. States of an atom: a) an atom under standard conditions; b) an atom being acted upon by an electric field; c) an atom being acted upon by a magnetic field.

The electromagnetic properties of materials are also dependent upon the number of electrons at the outer energy level, which is well known to be eight in number. The fewer the number of electrons at the outer energy level, and the greater the number of energy levels, the better the conductivity of the material.

The number of electrons at the outer energy level of a dielectric is so big that their orbits get deformed and go beyond the dielectric boundaries. They are more easily excited during friction and generate a surface charge (static electricity) (see Fig. 4.3).

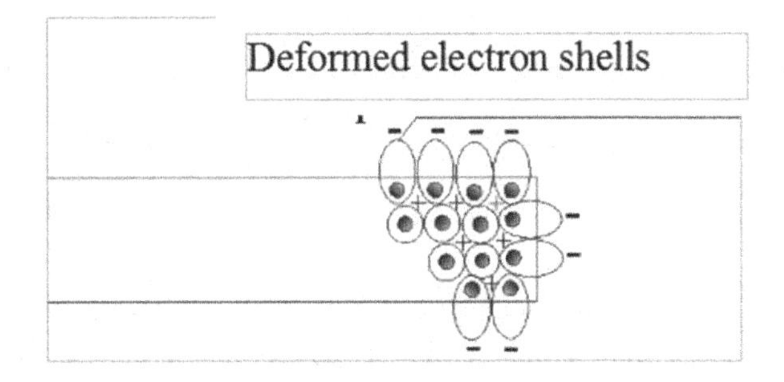

Fig. 4.3. Charge generation at the dielectric surface.

The electric current is characterized by the properties of the electromagnetic field because the electric field cannot exist without the magnetic field, and vice versa.

However, one of the basic characteristics specific to the electric current is the intensity of the electrical field. In fact, this is its primary characteristic. We assume that the intensity is a value of the deformation of the photon field in the upper layer of the conductor, which results in a vector consisting of electrical and magnetic components.

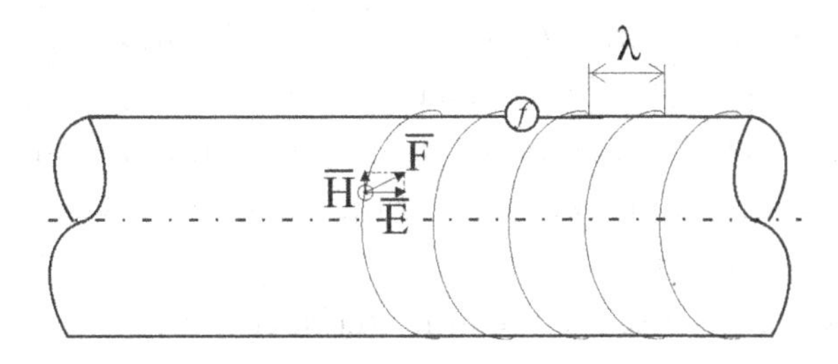

Fig. 4.4. The diagram of the electromagnetic wave propagation in the upper layer of the conductor

The deformation of the photon field provides for the transfer of waves according to Figure 4.4, and initiates polarization of the photon field according to Figure 4.5.

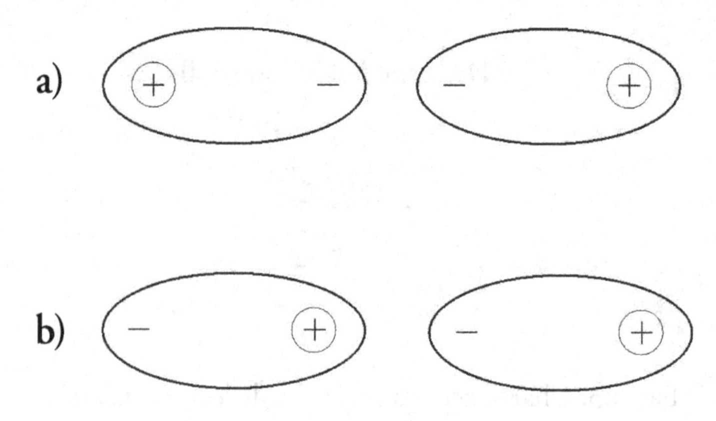

Fig. 4.5. The diagram of photon field polarization:
a) photons gravitate; b) photons repel.

An experiment performed in the USA provides a good example of the polarization of photons "In the course of experiments with a laser beam, physicists from several scientific institutes (including the Massachusetts Institute of Technology and the California Institute of Technology) have discovered the specific interaction of photons which makes it possible to talk about the formation of an earlier, previously unknown state of matter. The journal *Nature* published these physicists' work. Their paper describes the creation of a non-linear, quantum medium in which photons spread as massive and attracting particles. PhysOrg provides a brief summary of the experiment (with a reference to Harvard University).[9]

The researchers used a vacuum chamber with rubidium atoms. The temperature of the atoms was lowered close to absolute zero by so-called laser cooling. Weak laser pulses were then sent through the cloud of rubidium atoms. Observations indicated that the distribution of photons goes on with a lower speed in such an unusual medium (which is, according to the scientists, predictable because the speed of light lowers in any substance). Photons also gravitate towards each other. The motion of a pair of photons through the rubidium cloud can be described as the movements of a diatomic molecule: such effects on light were predicted earlier in theory, but had not been observed in practice.

New ideas of photon field properties mean that it is possible to explain a skin-effect in the upper layer of the conductor. This is known in theoretical physics as the propagation of electromagnetic waves in the upper layer of the conductor, but not in the conductor itself. At higher frequencies, the depth of the electromagnetic wave penetration is less; at super-high frequencies, it is a micron deep.

It is difficult to explain this phenomenon using current theoretical views. However, explanation becomes easier if we adopt the idea of a spiral distribution of an electric current in the boundary layer of the conductor.

Figures 4.4 and 4.5 indicate the universal character of energy distribution in space; these diagrams help clarify many complex phenomena and processes.

Thus, if distance λ (Fig. 4.4) decreases, the expectation is that this is caused by the increase of the frequency of the electric current and by the increase in its density, and, hence, transferred energy. The diagram shows this relationship in simplified form.

A significant increase of the frequency of the current might also cause an ordinary conductor to develop inductance. This will happen according to the following physical relations:

$$\frac{di}{dt} = L, \tag{4.1}$$

Where:
di – current change, A;
dt – time change, s;
L – inductance, H.

According to the physical relation (4.1), the higher the value of L, the higher the total value of the resistance of the conductor will be. If the intensity of the electric current passing through the conductor decreases, the voltage at this section increases. It is appropriate here to draw an analogy between the frequency of the loops of the electromagnetic wave propagation along the conductor surface (which depends upon the angle of the resulting vector of photon deformation) and the pitch of a screw. If resistance of the section of the conductor increases, the angle of the resulting vector towards the direction of the electromagnetic wave propagation also increases. At the same time the intensity of the current decreases. But the voltage increases proportionately to the unit of the conductor's overall length. By analogy with a screw, as the screw is driven into material, greater rotational force is required as resistance to it increases. If a screw with a smaller pitch is used, the value of the force of rotation remains the same (similar to voltage), but the longitudinal speed of the motion of turning decreases (similar to current).

The screw analogy also helps to illustrate the existence of properties that distinguish the electric current from other kinds of flows and similar mechanical movements. With screw-driving, changing the properties of the medium will lead to a change in three primary parameters related to the movement of the screw: rotational force, pitch, and the value of axis movement. Regarding the electromagnetic wave propagation in the conductor, changing the properties of the conductor causes a change of the three primary parameters: the voltage at the section of the conductor, the angle of the resulting vector of photon deformation, and the intensity of the current. This result confirms that if resistance to the wave propagation at the section of the conductor increases, then there will be a non-recoverable loss of energy for the electromagnetic wave. Thus it may be concluded that the electrical resistance of the conductor is a function of the relation between the values of the magnetic components and of the electrical components of photon deformation $R = f\left(\frac{H}{E}\right)$. It further relates to the degree of deformation of this photon, and also depends upon the density of electrons at the outer energy level of the conductor atom.

This definition of the electric current is true for a certain density of the electromagnetic wave energy. It also gives grounds for establishing characteristic differences between the electric resistance and other types of resistance, such as the hydraulic one, for example. It also helps to explain other, particular cases of the electric resistance of the conductor.

The suggested mechanism for the growth of electrical resistance in response to the increase of the magnetic component in total energy (distributed throughout the conductor of the electromagnetic wave) is common for all known, particular cases. Let's try to get to the bottom of this.

Electrical engineering teaches us that the winding of the conductor, which can ensure the power supply of a heavy-duty consumer when unwound, can lead to the conductor overheating and burning out. This effect is a result of additional resistance which occurs inside the wound-up conductor. The conductor then manifests magnetic properties as a result of the inductance generated in the coil. It is worth noting that the more frequent the windings of the coil of the conductor, the higher is its inductance. There is also a corresponding increase in magnetic properties and additional resistance. This additional resistance is generally thought possible due to the induction currents. (Experienced welders know about such phenomena and, unlike novices and amateurs, take the precaution of always unwinding a loading coil prior to the commencement of work to prevent its burn-out.)

However, we believe that a more accurate description of the mechanism for the formation of additional resistance is related to a significant dissipation and emission of part of the energy transferred by an electromagnetic wave. Ultimately, the emitted energy heats up the given wound section of the conductor.

When the electromagnetic wave transfers certain types of energy throughout the conductor, a specific density of energy is distributed along its surface. The increase of the magnetic component of photon deformation results in the active emission and dissipation of the energy of the electromagnetic wave, which is distributed throughout the given section of the conductor, and which heats it up.

Thus, according to the foregoing analysis, we can assume that both the electrical resistance of the material of the conductor, and the electrical resistance of the conductor itself, are directly related to the level of magnetism present.

This is the rationale for the increase of electrical resistance and energy loss of a given section of the conductor at higher frequencies of the current going through that section. In addition, it helps to explain the decrease in the depth of penetration of the electric wave, which is related to the growth of the magnetic component that pushes the electric wave out towards the surface.

Any review of accounts of the processes related to electromagnetic phenomena shows that they tend to assume the mutual attraction of like charges and mutual repulsion of unlike charges. However, these properties of electromagnetic phenomena are generally accepted as given conditions without any attempts to explain the phenomena.

At the outset of this analysis it should be noted that each elementary particle (electron, proton, neutron) is taken to be surrounded by an area of deformed space that is generated by electromagnetic waves. This area is considerably bigger than the particle. It is, therefore, reasonable to view the elementary particle as being within a cluster of electromagnetic disturbances in an area of deformed space.

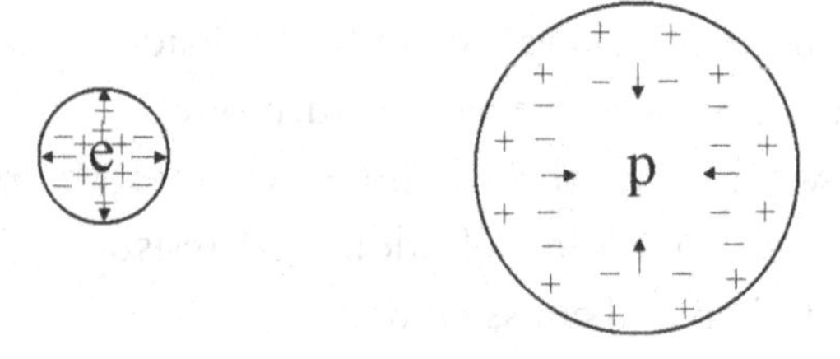

Fig. 4.6. The origin of the propagation of electromagnetic waves in electron (e) and proton (p)

Figure 4.6 shows ways for generating electromagnetic waves for elementary particles, with an electron and a proton as examples. Electromagnetic waves in the electron depicted here should be taken as propagating from the centre to the periphery along the spiral with a definite angle of expansion. Note that the electromagnetic wave extends beyond the electron.

Electromagnetic waves generated in the proton propagate from the periphery to the centre. They also extend beyond the proton with a certain angle of expansion.

To sum up, the electron is surrounded by a cluster of electromagnetic waves with vectors directed outwards, whereas the proton is surrounded by a cluster of electromagnetic waves with vectors directed inwards. If we consider the interactions of nearby electrons and protons, the vectors of action of their electromagnetic waves are directed in the same way and, in consequence, add up, which leads to the attraction of the particles. The opposite is the case with the interaction between elementary particles holding like charges: electron with electron and proton with proton.

This process becomes even clearer when we consider the formation of the polarization of the photon field as outlined in Figure 4.7.

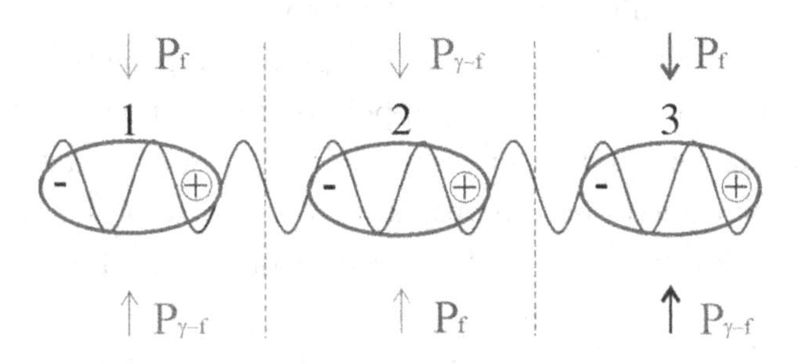

Fig. 4.7. Formation of the polarization of the photon field.
1, 2, 3 – polarized photons
$P_{\gamma\text{-}f}$ – pressure of the gamma-photon level
P_f – pressure of the non-polarized photon level

It is a generally accepted truism that unlike charges attract, whereas like charges repel. This postulate lends itself to a logical explanation.

Polarized photons have the shape of a spatial ellipsoid, and negative and positive charges are generated at the poles (see Fig. 4.7). Such a dipole, as with any other particle, will have its own frequency of oscillation and its own wave length and direction. According to our idea (see Fig. 4.7, photon 1), the electromagnetic wave will come out of the negative charge zone along the spiral with an expansion angle and an integer-multiple wavelength that allows it to pass through the positive charge zone. As the wave cannot propagate in one direction, it will propagate in the opposite direction with the same wavelength, with the same spiral and twisting direction, and with a similar expansion angle (the spiral expansion angle cannot be seen in the figure because of a small angle size).

Photons have the same structure, and identical processes take place in them. Therefore, there are multiple natural wavelengths with twisting directions and expansion angles. The potential pressure in the medium of the finer gamma-photon and non-polarized levels is constantly present. It is important to note, therefore, that the amount of energy of the potential pressure will decrease by the amount of kinetic energy of the oscillating particles in the areas where the electromagnetic wave propagates. Thus the polarized photons will fit into so-called "chains" under the influence of the energy of the potential pressure of the gamma-photon level and non-polarized photon level. They are formed by the composition of individual waves into the total spiral of the electromagnetic wave propagation. A spiral turn of one dipole wave is continued by the turn of the next dipole wave, while potential pressure presses them together.

We think that such a mechanism for the interaction of charged particles is manifested at various levels.

To fully understand the theory of the electric current, it is very important to analyse the functioning of heating and cooling of conductors.

Let's first review the processes going on in an atom of the conductor (Fe_{56}, for example) when it is heated.

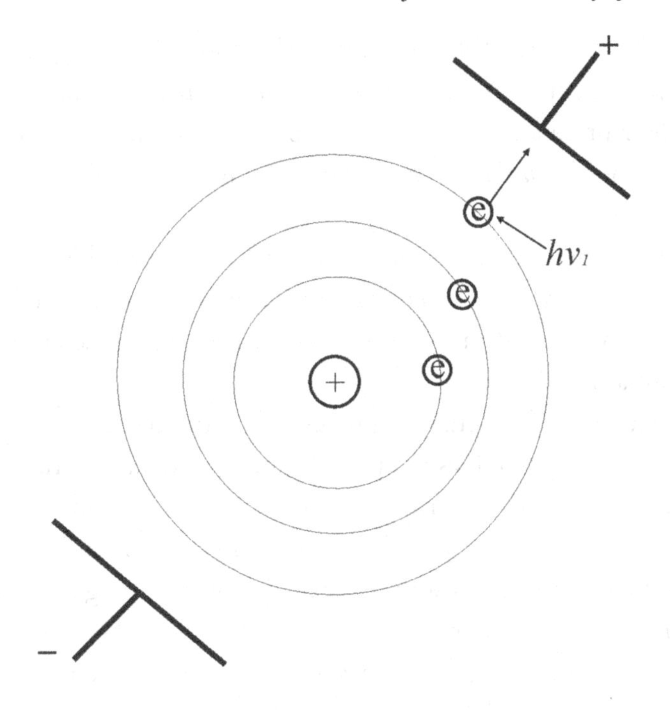

Fig. 4.8. The processes occurring in the conductor atom during heating.

Figure 4.8 shows the atom of the conductor located between positive and negative plates, and further heating.

It is a truism that heat energy ($h\nu$) is forced into the material of a conductor during heating. The total heat energy $\sum_i h\nu_i$, based on the oscillation frequency (ν_i), flowing to an atom is captured by electrons at the extreme energy level, which leads to a build-up of the total energy and hence the mass.

Consider, for example, an atom of iron. Before the inflow of heat energy to the atom, the mass of all 26 electrons at their orbits equals approximately 26 units of mass. After heating, the increased mass of the same 26 electrons might, supposedly, reach 27 integer units of mass. The atomic nucleus, which remains unchanged during the heating, can still retain electrons with a total of 26 integer units of mass. However, conditions occur when the nucleus can no longer retain the mass of 27 electrons. Conditions are created for the electrons of the external orbit to leave the boundaries of the atom. Free electrons with a negative charge then appear in the inter-atomic space and rush to the positive plate, where their presence can be detected by devices. This phenomenon is known as the *thermal electron emission*, and it consists of the formation of free electrons.

With further heating, more electrons leave their orbits. This leads to the destruction of the chemical, crystal bonds, and the rigid lattice is replaced by a liquid meltdown; material is transformed from a solid phase into a liquid phase.

In general, the heat-transfer process is a particular case of an energy transfer. Our research does not provide the close details of this process. However, it is worth mentioning because of the significance of the presence of the photon field. Taking the solar system as an example, it is well known that the greatest number of photons with high temperature (energy) surround the Sun, which actually produces them. The density of photons in interplanetary space is relatively low. The density increases close to the surface of the planets, such as the Earth, for the reasons mentioned. It follows that an electromagnetic wave carrying energy has no obstacles to its motion in interplanetary space. Moreover, the wave refraction coefficient in this medium is close to unity. This situation changes close to the surface of a planet. The density of the photon flow increases as the coefficient of wave reflection and refraction rises, which leads to the heating of the analysed area.

When the heat energy decreases, the material is transformed from the liquid phase into the solid phase, in the reverse order. Therefore, we will not describe this process separately.

Of more importance is an analysis of the phenomena that accompany a sharp drop in the temperature of the conductor.

Given that the cooling of materials, including conductors, results in the reduction of their linear dimensions, as the temperature decreases so does the distance in the inter-atomic and inter-nuclear spaces of the conductor – and this reduces the linear dimensions of the conductor. The positively charged nuclei of atoms start to interact with the internal electron clouds of neighbouring atoms. They form the so-called electron clouds, made up of common electron pairs (see Fig. 4.9).

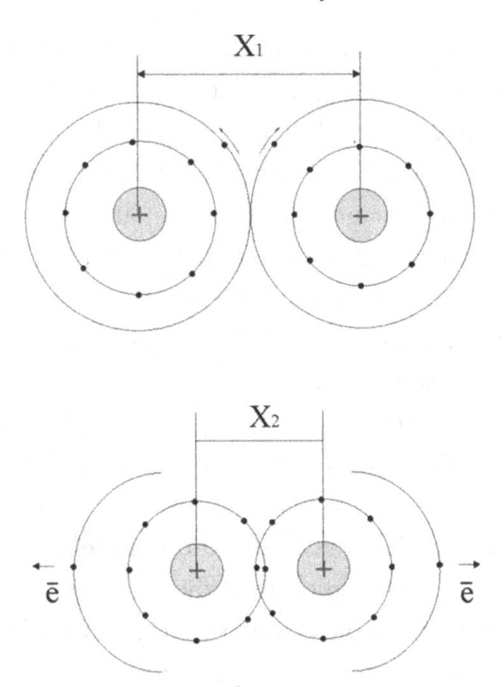

Fig. 4.9. The structural change of bonds in metal during cooling.

All chemical reactions involve the interaction of electrons at the outer electronic energy levels. As the temperature goes down, there is a moment when chemical reactions stop. This is a symptom of the destruction of the outer electronic energy levels. At the same time, inner electronic energy levels remain unchanged.

When the temperature of the conductor drops significantly, electrons at the outer energy level of the metal atom break away from their orbits all at once. These free electrons then start to participate in the process of electrical conductivity. The moment when the electrons simultaneously break away is known as the "Curie point", which implies that a given metal achieves a state of superconductivity.

4.2 The nature of magnetism

That the processes of electricity and magnetism are integral to each other is axiomatic. It is also true that a change in the magnetic field results in the occurrence of an electrical field, and vice versa – the electrical field creates the magnetic field. There is, therefore, no doubt about the interactive effects of magnetism and electricity.

An explanation of this sort is more difficult when considering the effects of a permanent magnet. In this case, all that is obvious is that the magnetic field is caused by the internal structure of the magnet. For example, if a big, permanent magnet having a north and south pole is broken into several pieces, any single piece, regardless of its size, will still have north and south poles.

Taking this fact into account, we suggest the following explanatory hypothesis. An external source of energy is used to form a permanent magnet. The source deforms, first, the electronic clouds, and, second, the nuclei of the material. The deformation of the atomic nuclei of the material is the source of magnetism of the future permanent magnet.

To put it figuratively, the external energy pumped into the nuclei of atoms stays inside the putative magnet for a prolonged period. The nucleus is a peculiar kind of energy accumulator.

Special devices can often record the electrical and magnetic characteristics of the electromagnetic field, which enables the properties of the permanent magnet to be attributed to its material. However, it is impossible to record the electromagnetic oscillations caused by the deformed atoms of such material, because the oscillation frequencies for both electric and magnetic fields of a permanent magnet are so high that it is impossible to distinguish even the separate harmonics of such oscillations.

Given this background, it is nonetheless possible to describe the chain of interactions within a permanent magnet. The passage of the electrical current through the material deforms the outer electronic clouds of the atom. The nucleus is then deformed and polarized. The nucleus then acts as a source for the magnetic field of the magnet later on. This process is similar to the accumulation of energy in a battery, and reminds one of the winding and unwinding of a spring.

Thus, all electromagnetic processes of the permanent magnet occur in resonance modes from within the deformed and polarized atoms of the material itself.

The long period of operation – 50 years and more – of permanent magnets in generators and engines is an indirect proof of this explanation.

In the light of this explanation, it is not unreasonable to analyse the magnetic field of the Earth in similar terms. Currently the Earth is regarded as acting like a huge permanent magnet with its corresponding magnetic field. However, we see it somewhat differently, and have the following, alternative logical explanations.

It is well known that the Earth, like many other planets, revolves around an axis that passes through its North and South Poles. The Earth is a conductor, and the magnetic component of its electromagnetic radiation creates lines of flux around it. If we take the Earth as a rotor which rotates in the field of the magnetic flux lines, then an electrical current should originate inside it. The direction of the current will be opposite to the direction of the Earth's rotation. Since electricity and magnetism are integral to each other, the occurrence of the magnetic field is self-evident. Moreover, the electromagnetic pulses are present throughout the process of the synthesis of matter from the photons in the interior of the Earth (as with a nuclear explosion, but the process takes much longer). In accordance with the right-hand rule, the vector of the magnetic field direction will approximately coincide with the Earth's rotational axis.

Continued electromagnetic pulses create the magnetic field of the Earth.

Therefore, it can be assumed that as long as the Earth rotates around its axis, and as long as there is synthesis of matter from the activity of the photons, the Earth will have its magnetic field. Furthermore, being a conductor is one of the primary conditions for the existence of the magnetic field of the Earth. After all, all the planets that have cooled down are dielectrics, with all the consequences that ensue.

The peculiarities of the Earth's shape should also be noted. The Earth is well known to be oblate. The increase of the Earth's diameter at the equator can possibly be explained by the centrifugal effect, which is at its maximum there. However, the oblate form is difficult to explain using current theories. Besides stretching along the equator, the Earth is exposed to the influence of forces pressing from the poles. These forces are due to the pressure of the gamma-photon level, the photon level and, even to some extent, air.

When the Earth rotates on its axis, different points on the Earth travel with different linear speeds.

The maximum linear speed occurs at the equator, whereas the minimum speed, almost zero, occurs at the North and South Poles.

The potential pressures of the gamma-photon level, the photon level, and the air are at their fullest at the poles. On other parts of the Earth's surface, the pressure will be considerably less, in accordance with Bernoulli's principle: that the potential pressure decreases in certain surface areas if the medium flow increases there (in this case, the increase of the linear speed of certain surface points). This implies that there will be minimum potential pressure at the equator and maximum potential pressure at the poles. The product of pressure by area equals the applied force of the pressure.

Apart from the poles of the Earth, it is worth mentioning the reasons for "the polar lights", which are connected with the magnetic lines of flux. The electromagnetic waves arriving at the Earth from the cosmos are blocked by magnetic lines of flux across the largest part of its surface. It happens because the waves approaching the lines are almost at right angles to them. However, at the poles, practically nothing is in the way of such electromagnetic waves, because their direction of travel almost coincides with the direction of the Earth's magnetic lines. The electromagnetic waves that come close to the Earth's surface near the poles interact with moisture in the atmosphere and result in the formation of colourful polar lights.

4.3 The mathematical representation of the magnetic field

The magnetic field is represented by two distinct vector fields denoted by the symbols H and B.

H refers to the magnetic field strength, and B to the magnetic flux density. The term "magnetic field" is used for both vector fields, though historically it referred mainly to H.

The density of the magnetic flux, B, is a primary characteristic of the magnetic field. Firstly, it determines the forces that influence the charges; secondly, vectors B and E are actually the components of the electromagnetic field tensor. Similarly, a single tensor comprises H and D, the electric displacement field. In its turn, the division of the electromagnetic field into the electric and the magnetic fields is conditional, and depends upon the chosen reference system. Therefore, vectors B and E should be reviewed together.

However, in a vacuum, and hence at a fundamentally microscopic level, H and B coincide (in the SI system calibrated to the conditional constant multiplier, but in the CGS system, completely). Thus, the researchers, especially those who don't use an SI system, can choose H and B at their discretion when defining the magnetic field. At the same time, those who use the SI system repeatedly favour vector B – if for no other reason than that it is used to directly express the Lorentz force.

4.4 The magnetic properties of substances

As noted above, from a fundamental point of view the magnetic field can be created, and therefore weakened or strengthened, by an alternating electrical field – that is, electric currents in the form of a flow of charged particles or magnetic moments of such particles.

The specific microscopic structure and the properties of different substances, as well as their mixtures such as alloys, aggregate conditions, crystal modifications, etc., bring about various responses to the action of an external magnetic field at the macroscopic level. For example, these factors might weaken or strengthen the magnetic field by varying degrees.

Based on their magnetic properties, the materials, and, more generally, their media, are divided into the following main groups:

Anti-ferromagnetic materials are the materials with an anti-ferromagnetic order of magnetic moments of atoms or ions; the magnetic moments of substances are equal and directed in opposite directions.

Diamagnetic materials create a magnetic field in opposition to an externally applied field.

Paramagnetic materials are materials which get magnetised in the direction of the externally applied magnetic field.

Ferromagnetic materials are the materials which establish the remote ferromagnetic order of magnetic moments after the critical temperature, the Curie point, has been achieved.

Ferrimagnetic materials have magnetic moments which are not equal and are directed in opposite directions.

These groups of materials usually include solid or liquid materials and gases. The interaction of the superconductors and the plasma with the magnetic field differs considerably.

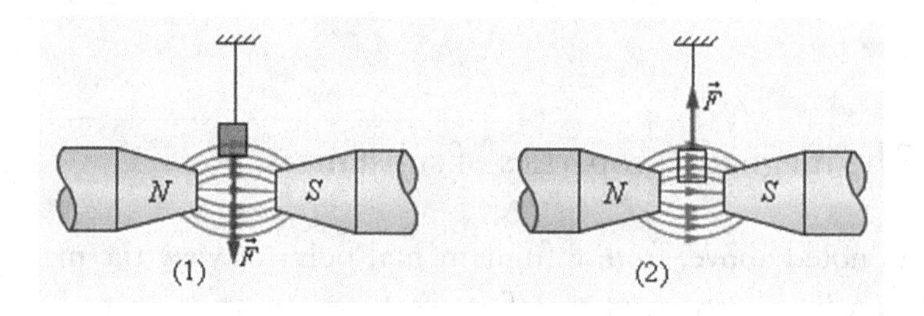

Fig. 4.10. Paramagnetic (1) and diamagnetic (2) material inside the inhomogeneous magnetic field.

4.5 The units of measurement

In Le Système International d'Unités (SI) units, B is measured in teslas; in Centimetre–Gram–Second Système (CGS) units, it is measured in gauss.

The H-field is measured in amperes per metre (A/m) in SI units, and in oersteds in CGS units. Since oersted and gauss are identical, the distinction between them is purely terminological.[10]

4.6 The electromagnetic nature of light; the plane wave in a dielectric material

In this section we will consider one of the simplest questions in the theory of electromagnetic waves: the propagation of the plane monochromatic waves in homogeneous dielectrics.

A wave shall be determined as a plane wave if its vectors have the same value at any time in all points of any plane perpendicular to the propagation of the wave. In other words, if we take the *z* axis as running along the direction of wave propagation, then vectors *E* and *H* of the field of the wave plane will depend upon the *z* coordinate alone, not upon coordinates *x* and *y*. The study of plane waves has a certain value from the physical point of view because it is possible to regard a definite section of a spherical wave emitted by an oscillator as a plane, when allowing for a degree of approximation.

A wave is monochromatic if the wave field is a harmonic – that is, a sinusoidal function of time. Note that "monochromatic" derives from the Greek meaning "one colour". It is the language of optics. Therefore, complex expressions of the field vectors of a monochromatic plane wave shall look as follows:

$$E = A(z)e^{i\omega t};\ H = B(z)e^{i\omega t} \qquad (4.2)$$

where the complex vectors A(z) and B(z) depend only upon the coordinate *z*. Of course, only the real component of these expressions has a direct physical value.

Let's assume that the dielectric in question is homogeneous and has no free electric charges (p = 0). By setting in the Maxwell's equation (I) j = 0 we have:

$$\frac{\varepsilon}{c}\frac{\mathbf{dE}}{\mathbf{dt}} = \text{rot } H \qquad (4.3)$$

by differentiating the expression with respect to time and by introducing value $\partial H/\partial t$ from equation (4.4):

$$\frac{\mu}{c}\frac{\partial H}{\partial t} = \text{rot E} \qquad (4.4)$$

based on this equation we obtain

$$\frac{\varepsilon}{c}\frac{\partial^2 E}{\partial t^2} = \text{rot}\frac{\partial H}{\partial t} = -\frac{c}{\mu}\text{rot E} =$$

$$= -\frac{c}{\mu}(\text{grad div E}) - \nabla^2 E \quad (4.5)$$

Thus, if p — 0 and E = const, we get the following equation:

$$\text{Div D} = \varepsilon \text{ div E} = 0 \qquad (4.6)$$

so

$$\frac{\varepsilon\mu}{c^2}\cdot\frac{\partial^2 E}{\partial t^2} = \nabla^2\text{E}. \qquad (4.7)$$

As equations (4.3) and (4.4) are symmetrical with regard to E and H (up to a decimal place), we similarly obtain:

$$\frac{\varepsilon\mu}{c^2}\cdot\frac{\partial^2 H}{\partial t^2} = \nabla^2\text{H}. \qquad (4.8)$$

The correctness of equations (4.7) and (4.8) is limited only by the requirement for a homogeneous medium, absence of conduction currents, and free charges. In case of plane monochromatic waves, based on equation 4.2 these equations can be written as follows (reducing by $e^{i\omega t}$):

$$-\frac{\varepsilon\mu\omega^2}{c^2}A(z) = \nabla^2 A(z) = \frac{\partial^2 A(z)}{\partial z^2},$$

$$-\frac{\varepsilon\mu\omega^2}{c^2}B(z) = \nabla^2\text{B}(z) = \frac{\partial^2\text{B}(z)}{\partial z^2} \qquad (4.9)$$

or

$$\frac{\partial^2 A}{\partial z^2} + k^2 A = 0, \frac{\partial^2 B}{\partial z^2} + k^2 B = 0, \tag{4.10}$$

where we introduce a symbol

$$k^2 = \frac{\varepsilon\mu\omega^2}{c^2}, \ \omega\frac{\sqrt{\varepsilon\mu}}{c}. \tag{4.11}$$

As we know, the solutions for such equations appear as follows:

$$A = A_0 e^{-ikz} + A_0' e^{ikz},$$
$$B = B_0 e^{-ikz} + B_0' e^{ikz} \tag{4.12}$$

where A_0, A_0', $B_{0,}$ and B_0' are the arbitrary constants of integration. Introducing these expressions to equation 4.1 we obtain

$$E = A_0 e^{i(\omega t - kz)} + A_0' e^{i(\omega t + kz)}$$
$$H = B_0 e^{i(\omega t - kz)} + B_0' e^{i(\omega t + kz)} \tag{4.13}$$

The first terms of these expressions represent a wave which propagates in the positive direction along axis *z*; the second terms represent a wave which propagates in the opposite direction. It will not impair the general reasoning if we consider only one of the waves – for example, the first one – and assume that

$$E = A_0 e^{i(\omega t - kz)}, \ H = B_0 e^{i(\omega t - kz)} \tag{4.14}$$

A_0 and B_0 are amplitudes of vectors *E* and *H*. The amplitudes are independent of the coordinates. This means that the propagation of plane waves in dielectric material is not connected with the change of their intensity. Generally speaking, the amplitudes A_0 and B_0 are complex values.

According to equation (4.14), the speed of wave is ω/k, because at the moment of time t_0 the values of the field vectors in plane z = Z0 coincide with the values they had at the moment of time t_0 — 1 in plane z = z_0 — ω/k. This becomes apparent from the equality of the respective phases:

$$\omega t_0 - kz_0 = \omega(t_0 - 1) - k\left(z_0 - \frac{\omega}{k}\right) \tag{4.15}$$

According to equation (4.11) this speed is equal to

$$v = \frac{\omega}{k} = \frac{c}{\sqrt{\varepsilon\mu}} \tag{4.16}$$

Let's note that *k* value is ordinary connected with the wavelength λ; by introducing $\omega = 2\pi/T$ into equation 4.16, we obtain:

$$k = \frac{\omega}{v} = \frac{2\pi}{Tv} = \frac{2\pi}{\lambda} \tag{4.17}$$

Thus, *k* equals the number of waves that exist over a specified distance 2π cm. Therefore, it is called a wave number.

To simplify calculations further, let's note that according to equation (4.14), the differentiation of the plane wave vectors with regard to *z* comes down to multiplication by *ikk*. On the other hand, these vectors do not depend on *x* and *y*. Therefore, the symbolic multiplication of the vectors by the differential operator Nabla comes down to multiplication by a usual vector — *ikk*. Thus, as applied to these vectors:

$$\nabla = \mathrm{i}\frac{\partial}{\partial x} + \mathrm{j}\frac{\partial}{\partial y} + \mathrm{k}\frac{\partial}{\partial z} = -\mathrm{ikk} \tag{4.18}$$

(the unit vectors *i* and *k* along the axes of coordinates should not be confused with the unit imaginary number i and the wave number k).

If axis *z* does not coincide with the direction of the wave propagation, it is enough to replace *k* with a unitary vector *n* which coincides with this direction:

$$\nabla = -ikn \tag{4.19}$$

Based on this relation, the Maxwell's equations (4.5) and (4.6) take the following form:

$$\text{div B} = \nabla\mu\text{H} = -ik\mu n\text{H} = 0,$$
$$\text{div D} = \nabla\varepsilon\text{E} = -ik\varepsilon n\text{E} = 0,$$

whence it follows that vectors *E* and *H* are perpendicular to *n* – that is, perpendicular to the wave direction. *Thus, the plane electromagnetic waves, as well as spherical waves, are transverse waves.*

According to equation 4.14, the time differentiation of vectors *E* and *H* comes down to multiplication by ω. Thus, equation 4.4 can be presented as follows using equation 4.19:

$$\frac{i\omega\mu}{c}H = -\text{rot E} = -[\nabla\text{E}] = ik[\text{nE}]$$

By introducing *k* value from equation 4.11 and by dividing the equation by iω, μ/s, we obtain

$$\sqrt{\mu}\text{H} = \sqrt{\varepsilon}[nE] \tag{4.20}$$

It follows from this equation that, first, vectors *E* and *H* are mutually perpendicular; and, second, vectors *n*, *E*, and *H* form the right-handed system.

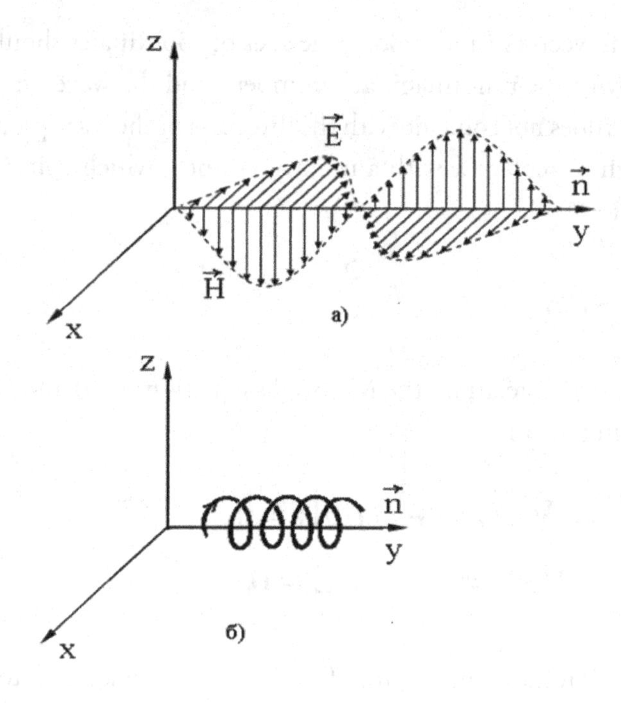

Fig. 4.11. The spiral wave.

Further, due to the fact that *n* and *E* are perpendicular, we obtain the following relation between the numeric values of vectors *E* and *H*:

$$\sqrt{\mu}H = \sqrt{\varepsilon}\left|[nE]\right| = \sqrt{\varepsilon}E \tag{4.21}$$

Thus, the relation of the numeric values of vectors *E* and *H* does not depend on time; the vectors have equal phases and are synchronous.

Before referring to the definition of the non-linear functions of the field vectors (energy, the Poynting vector, etc.), we need, first of all, to consider the real components of complex expressions (4.14). Therefore, vectors *E* and *H* will be regarded as real vectors in further formulas.

According to equation (4.21), the density of the magnetic energy in the field of the wave equals the density of the electric energy:

$$w_m = \frac{\mu H^2}{8\pi} = \frac{\varepsilon E^2}{8\pi} = w_e \tag{4.22}$$

so

$$w = w_m + w_m = \frac{\varepsilon E^2}{4\pi} = \frac{\mu H^2}{4\pi} \tag{4.23}$$

After review of Figure 4.6, it becomes clear that the direction of the Poynting vector

$$\mathrm{S} = \frac{\mathrm{c}}{4\pi}[\mathrm{EH}], \tag{4.24}$$

that is, the direction of the flow of energy in the wave coincides with the direction of its propagation. Because vectors E and H are perpendicular,

$$S = \frac{c}{4\pi} EH. \tag{4.25}$$

Expressing H in terms of E by means of equation (4.20), and using equations (4.22) and (4.15), we obtain

$$S = \frac{c\sqrt{\varepsilon}}{4\pi\sqrt{\mu}} E^2 = \frac{c}{\sqrt{\varepsilon\mu}} w = wv \tag{4.26}$$

whence

$$S\ \partial t = wv\ \partial t \tag{4.27}$$

Thus, the amount of energy S that flows in a certain time interval dt across a unit area perpendicular to vector S (that is, perpendicular to the wave direction) equals the amount of energy $wv\ dt$ which is contained in a cylinder with height $v\ dt$ adjacent to this area. From the physical standpoint it means that the energy flow velocity equals v – that is, it coincides with the tangential velocity of the wave.

4.7 The reflection and refraction of plane waves in dielectrics

In this section we will dwell upon the refraction and reflection of plane monochromatic waves at the interface surface of two homogeneous dielectrics 1 and 2. Let's set the dielectric constants as ε_1 and ε_2. For the sake of simplicity, let's accept $v_1 = v_2 = 1$. It will readily be seen that this assumption does not limit the general character of our reasoning for light waves. Under these conditions, according to equation (4.15) the speed of waves in the first and the second dielectrics will be equal to, respectively,

$$v_1 = \frac{c}{\sqrt{\varepsilon_1}} \text{ and } v_2 = \frac{c}{\sqrt{\varepsilon_2}} \tag{4.28}$$

Earlier we assumed that the direction of axis z is chosen in such a way as to coincide with the direction of the wave. Considering the totality of several waves with different directions (incident, reflected, and refracted waves), we should summarize the formulas of the previous paragraphs in cases where the coordinate axes are directed arbitrarily. Suppose the direction of a wave coincides with the direction of a unitary vector n, which forms angles α, β, γ with coordinate axes x, y, and z (see Fig. 4.12).

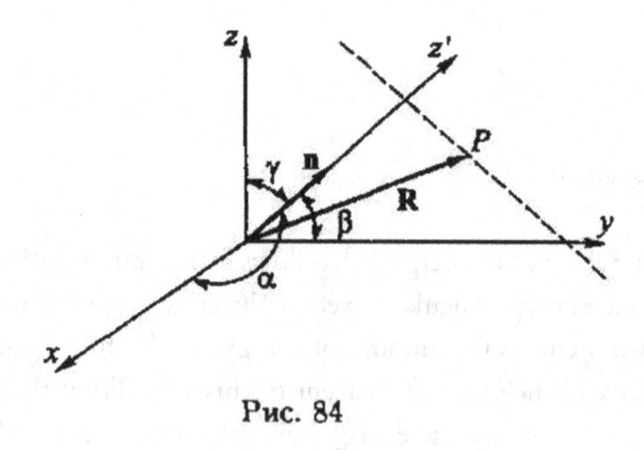

Fig. 4.12. Wave direction.

In the Cartesian coordinate system x', y', z', where z' coincides with *n*, the phase of the wave at a point *P* shall be defined by the expression below and according to equation (4.14)

$$\omega t - kz', \tag{4.29}$$

where z' is the coordinate of the point *P*. This coordinate is equal to the projection of radius-vector *R* of the point *P* on the direction *n* of the axis z', therefore

$$z' = nR = x\cos\alpha + y\cos\beta + z\cos\gamma, \tag{4.30}$$

where *x*, *y*, *z* are the coordinates of the same point *P* in the original coordinates system. By introducing this into equation (4.14) we obtain the desired expressions of the field vectors of the plane wave, which propagates in an arbitrary direction *n*:

$$E = E_0 e^{i(\omega t - knR)}, \quad H = H_0 e^{i(\omega t - knR)} \tag{4.31}$$

where E_0 and H_0 are the constant complex amplitudes of the respective vectors.

After these preliminary remarks, let's proceed with finding a solution to our problem. Let the plane wave (4.31) which propagates through medium 1 in direction *n* be incident on the plane surface of interface of media 1 and 2. Upon penetration into medium 2 the wave will propagate with a different speed [$v_2 = v/h$] and, as we shall see, in a direction different from *n*.

To define the amplitudes, directions, and phases of this refraction wave, it is enough to fulfil the aforementioned boundary conditions at the interface surface.

It turns out that such conditions can be fulfilled only if it is granted that there is a third, so-called reflected wave, which propagates in medium 1 as the incident wave does, but in the direction from the interface surface.

Let's set the complex vectors of the field of the incident wave E and H; the vectors of the reflected wave W and H_r; and, finally, the vectors of the refracted wave Eg and Hg; and let's assume that according to equation (4.31)

$$E = E_0 e^{i(\omega t - knR)}, \quad E^r = E_0^r e^{i(\omega_r t - k_r n_r R)},$$

$$E^g = E_0^g e^{i(\omega_g - k_g R)} \tag{4.32}$$

We are not going to put down similar expressions for H, $H\varepsilon$ and Hg. The first two waves propagate through medium 1, so the resultant field strength in this medium will be

$$E_1 = E + E^r \tag{4.33}$$

whereas in medium 2, it will be

$$E_2 = E^g \tag{4.34}$$

Let's consider any boundary condition at the interface surface – for example, the continuity of tangential components of vector E. In the case under consideration it will be as follows:

$$E_t + E_t^r = E_t^r \tag{4.35}$$

That is,

$$E_{0t} e^{i(\omega t - knR)} + E_{0t}^r e^{i(\omega_r t - k_r n_r R)} = E_{0t}^g e^{i(\omega_g t - k_g n_g R)} \tag{4.36}$$

To ensure the fulfilment of such conditions at any given time t, it is required, first and foremost, to make sure

$$\omega = \omega_r = \omega_g \tag{4.37}$$

Indeed, the equation (4.36) takes the form

$$ac^{i\omega t}+bc^{i\omega_r t}=ce^{i\omega_g t} \tag{4.38}$$

where *a*, *b*, and *c* do not depend upon time. By differentiating this equality with respect to *t*, we obtain

$$\omega ae^{i\omega t}+\omega_r be^{i\omega_r t}=\omega_g ce^{i\omega_g t} \tag{4.39}$$

By excluding $ce^{i\omega_g t}$ from the two last equations, we obtain

$$a\left(\omega-\omega_g\right)e^{i\omega t}=b\left(\omega_g-\omega_r\right)e^{i\omega_r t} \tag{4.40}$$

which is possible only if $\omega = \omega_g$. By excluding $e^{\omega_r t}$ from these equations, we find out that $\omega = \omega_g$. It follows that $\omega = \omega_r = \omega_g$ – that is, the wave frequency does not change at its reflection and refraction.

Similarly, we find out that the following equalities will be satisfied:

$$knR'=k_r n_r R'=k_g n_g R' \tag{4.41}$$

where R' is an arbitrary vector lying on the interface surface. Indeed, the radius-vector *R* of an arbitrary point on the surface interface can be written in the form

$$R=R_0+R' \tag{4.42}$$

where *R'* lies on the surface of the media interface, whereas *Ro* is a radius-vector of any arbitrary fixed point on this surface.

Therefore, the boundary condition can be presented as follows

$$a'e^{-iknR'}+b'e^{-ik_r n_r R'}=c'e^{-ik_g n_g R'} \tag{4.43}$$

where values a', *b'*, and *c'* do not depend upon the coordinates of vector *R'*. The further proving of equation (4.41) is similar to the proving of equation (4.37).

For convenience, let's move to coordinate expressions for the next calculations. Let axis *z* be perpendicular to the interface plane of media 1 and 2. Then vector *R'* belonging to this plane will be perpendicular to axis *z*; and, based on equation (4.30) it is possible to write equation (4.41) as follows:

$$k(x\cos\alpha + y\cos\beta) \\ = k_r(x\cos\alpha_r + y\cos\beta_r) \\ - k_g(x\cos\alpha_g + y\cos\beta_g) \tag{4.44}$$

where *x* and *y* are components of vector *R'*, and α, *β*; α_g, β_g; α_g, n_g are respective angles of vectors *n*, n_r and n_g with axes *x y z*. For simplicity, let's take axes *x* and *y* so that the direction of the propagation of the incident wave *n* lies in the plane *xz* (see Fig. 4.13).

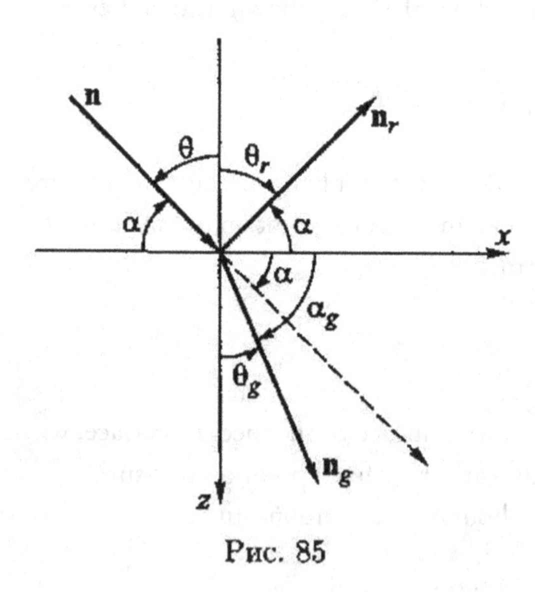

Fig. 4.13. The direction of incident wave propagation.

In this case $\cos\beta = 0$. Since the presented equation will be fulfilled for all points of the interface plane – that is, at any values of x and y – it immediately follows that:

$$\cos\beta_r = \cos\beta_g = 0 \tag{4.45}$$

and

$$k \cos \alpha = k_r \cos \alpha_r = k_g \cos \alpha_g \tag{4.46}$$

The first equation means that the direction of the reflected and refracted waves n_r и n_g lie in plane *xz* (that is, in the wave incident plane).

Let's take into account that the incident and reflected waves propagate in medium 1, while the refracted waves in medium 2; therefore, according to equations (4.16) and (4.17)

$$k = \frac{\omega}{v_1},\ k_1 = \frac{\omega}{v_1} = k,\ k_g = \frac{\omega}{v_2}. \tag{4.47}$$

It follows that, first, $\alpha_g = \alpha$. By introducing the incident and reflection angles n and n_r (as is apparent from Figure 4.12, $\boldsymbol{\theta} = \pi/2 — \alpha$ and $\boldsymbol{\theta}_r = \pi/2 — \alpha_g$), we can say that the incident angle в is equal to the reflection angle n_r.

Further, as $\cos\alpha$ and $\cos\alpha_g$ have the same signs, the directions of the incident and refraction waves shall lie in the same quadrant of the plane xz (see Fig. 4.13).

By introducing the refraction angle $\boldsymbol{\theta}_g = \pi/2 – \alpha_g$ and noting that $\cos\alpha = \sin\boldsymbol{\theta}$ and $\cos\alpha_g = \sin\boldsymbol{\theta}_g$, we obtain:

$\sin\boldsymbol{\theta}/v_1 = \sin\boldsymbol{\theta}_g/v_2$, (4.48)

or

$$\frac{\boldsymbol{sin\theta}}{\boldsymbol{sin\theta_g}} = \frac{\boldsymbol{v_1}}{\boldsymbol{v_2}} \tag{4.49}$$

Thus, the relation of sines of the incident angle and the refraction angle is a constant value, which depends only upon properties of the adjacent media 1 and 2. As we know, this relation denoted by n_{12} is called a refractive index of medium 2 with regard to medium 1. Based on equation (4.28) we can write out

$$n_{12} = \frac{v_1}{v_2} = \sqrt{\frac{\varepsilon_2}{\varepsilon_1}} \tag{4.50}$$

Thus, all the laws of geometry for the refraction and reflection of the electromagnetic waves follow from the fact that the linear conditions of the type (4.36) exist at the interface of the two media. These laws coincide with the respective laws for light waves. Further detailed study of such boundary conditions makes it possible to determine the relation between the squares of the amplitudes of the incident wave and the reflected/refracted waves. Such relations prove to be identical with the so-called Fresnel equations that determine the relative intensity of reflected/refracted light based on the refraction coefficient, the incident angle, and the polarization of the incident light. The development of the experimentally confirmed Fresnel equations from the general provisions of electrodynamics is one of the most important proofs of the electromagnetic nature of light. Not having enough place to provide a full review, we will limit ourselves to a review of the simplest case: the normal incidence of the plane wave to the interface surface of dielectrics z = 0.

It is possible to separate any monochromatic plane wave into the totality of two linearly polarized waves which can be reviewed individually. Let the electric vector of the incident wave be directed along axis x:

$$E_x = E_0 e^{i(\omega t - kz)}, \quad E_y = E_z = 0 \tag{4.51}$$

Then, according to (4.20) the magnetic vector of the incident wave will be directed along axis y and will be

$$H_x = H_z = 0, \quad H_y = \sqrt{\varepsilon_1} E_0 e^{i(\omega t - kz)} \tag{4.52}$$

Based on the laws of reflection and refraction and equation (4.49), the refracted and the reflected waves will be directed respectively along and against axis z. As it is easy to verify by looking at the boundary conditions, the electric vectors of these waves, as well as for the incident wave, are directed along axis x, whereas the magnetic vectors are directed along the axis y. Thus the non-zero components of these vectors are equal to

$$E_x^r = E_0^r e^{i(\omega t + kz)}, \quad H_y^r = -\sqrt{\varepsilon_1} E_0^r e^{i(\omega t + kz)} \tag{4.53}$$

$$E_x^g = E_0^g e^{i(\omega t - k_g z)}, \quad H_y^g = \sqrt{\varepsilon_2} E_0^g e^{i(\omega t - k_g z)} \tag{4.54}$$

The minus sign is put in front of the expression for H_y^r because the reflected wave propagates opposite to axis z; see equation (4.50).

Let's introduce the expressions for E and H into the boundary conditions. The condition (4.36) for continuity of tangential components of the vector E at the interface plane $z = 0$ after reducing by $e^{i\omega t}$ results in

$$E_0 + E_0^r = E_0^g \tag{4.55}$$

Consequently, the condition for the continuity of tangential components of the vector H results in

$$\sqrt{\varepsilon_1} E_0 - \sqrt{\varepsilon_1} E_0^r = \sqrt{\varepsilon_2} E_0^g \tag{4.56}$$

Normal components of the vectors of the field are equal to zero; therefore, the related boundary conditions shall be identically satisfied.

Solving the given equations and using the relation (4.50), we obtain

$$E_0^r = \frac{1 - n_{12}}{1 + n_{12}} E_0, \quad E_0^g = \frac{2}{1 + n_{12}} E_0 \tag{4.57}$$

Passing over to the real parts of complex expressions and regarding E_0 as being real, we can write out

$$E_x^g = \frac{2}{1+n_{12}} E_0 \cos\left(\omega t - k_g z\right), \tag{4.58}$$

$$E_x = E_0 \cos\left(\omega t - kz\right),$$

$$E_x^r = \frac{1-n_{12}}{1+n_{12}} E_0 \cos\left(\omega t + kz\right) \tag{4.59}$$

By means of (4.50) it is easy to verify that these expressions satisfy the law of conservation of energy – that is, an average flow of energy in the incident wave is equal to the amount of average flows of energy in the refracted and reflected waves:

$$\overline{S} = \overline{S^r} + \overline{S^g} \tag{4.60}$$

The reflection coefficient *r* of the surface at the interface between two media shall be identified as the relation of the reflected wave intensity and the incident wave intensity; in other words, the relation of average flows of energy in the reflected and incident waves:

$$r = \frac{\overline{S^r}}{\overline{S}} = \left(\frac{1-n_{12}}{1+n_{12}}\right)^2 \tag{4.61}$$

When $n_{12} = 1$ (that is, when $V_1 = V_2$), there is no reflection.

Thus, we have identified all the parameters that characterize the reflected and refracted waves in a case of the normal incidence of an original wave on the interface surface. The expressions we derived coincide with the respective Fresnel equations.

Besides the development of the light reflection and refraction laws, which were known before the discovery of light's electromagnetic nature, the theory we suggest helps to establish the direct relation between the reflection coefficient, light *n* and dielectric permittivity ε [formula (4.50)]. In particular, the refraction coefficient of a dielectric with regard to a vacuum (for which ε = 1) will be equal to

$$n = \sqrt{\varepsilon} \tag{4.62}$$

For some dielectrics (for example, air, carbon monoxide (CO), benzene), formula (4.62) meets the experimental test. However, for many substances this formula does not correspond to the measurements:[11]

Table 4.1 Parameters of substances

Substance	n	$\sqrt{\varepsilon}$
water	1.33	9.0
methyl alcohol	1.34	5.7
ethyl alcohol	1.36	5.0

Moreover, the fact of light dispersion – that is, the dependence of the refraction coefficient on the wavelength – argues against formulas (4.50) and (4.62), according to which n shall be constant for all electromagnetic waves.

Thus, Maxwell's phenomenological theory of the macroscopic field leads, generally speaking, to incorrect values for the refraction index. However, this contradiction is easily resolved from the point of view of the electron theory of the macroscopic field. Indeed, when we derived equations of the macroscopic field in Chapter 2, we likened molecules of the dielectric with electric dipoles. If these dipoles are quasi-elastic, they will have the natural period of oscillations. If this period is close to the period of light waves, the amplitude of oscillations of the dipole charges excited by the alternating field of the light wave will depend to a considerable extent upon the amplitude of the electric field of wave E, and also upon the period (or length) of the light wave (resonance). Therefore, the amplitude of the varying polarization vector of dielectric P, as well as the amplitude of the electric displacement field vector D = εE will depend to a considerable extent upon the period or length of the light wave. Thus, considering the peculiarities of the microscopic structure of dielectrics, we are able to determine a certain dependence between the value of dielectric permittivity and the wavelength – and, therefore, according to equation (4.62), to clarify the light dispersion. By introducing to equation (4.62) the value of dielectric permittivity measured in the constant or variable field, it is obvious that we can determine the value of the refraction index only for comparatively long electromagnetic waves. The period of such waves will be much longer as compared with the natural period of oscillations of the dielectric dipoles. For such waves, formula (4.62) actually meets the experimental test.

The situation is somewhat different if we liken molecules of the dielectric to a solid dipole. A major part of the polarization of such dielectrics reduces itself to the rotation of axes of dipoles (that is, the molecules) in the direction of the field. In the quick-changing fields, the axes of the molecules of the dielectrics that have a certain moment of inertia do not follow the fast changes of the field directions. As a result, the dielectric gets polarized to a lesser extent when compared to a constant electric field of the same strength. Therefore, the value of dielectric permittivity (and, thus, the value of the refraction index) in the dielectrics of this category (water, spirits, etc.) rapidly declines as the period of the field oscillations decreases. Such a decline occurs at frequencies which are much lower than the frequencies of the natural oscillations of the quasi-elastic molecular dipoles (for example, in some spirits it occurs at radio frequencies).

The same explanation is true for the fact mentioned at the beginning of the above paragraph: that it is possible to consider $\mu = 1$ for the study of the light waves. Indeed, the susceptibility of diamagnetic substances is so insignificant that the difference of μ from unity can be neglected if the light speed is determined by formula $v = c/\varepsilon\mu$. The mechanism of paramagnetic magnetizing is similar to the mechanism of the polarization of the dielectric with solid dipoles. Therefore, the paramagnetic susceptibility k becomes practically equal to zero, whereas permittivity μ becomes equal to unity at frequencies that are considerably less than the frequency of the visual light. Finally, the magnetizing of ferromagnetic materials is specified by the change in direction of the magnetizing of certain saturated Weiss domains, as well as the change in the size of such domains. These processes also lag behind the changes of the light-wave field. Therefore, from the optical point of view, the ferromagnetic materials do not differ much from the diamagnetic materials.

4.8 The propagation of waves in the conductive medium; the reflection of light from a metal surface

The electromagnetic waves propagate through dielectric material without any damping. However, in good conductors –metal, for example – the electromagnetic waves damp so quickly that even the thinnest layers of metal are non-transparent for the waves. This is, of course, explained by the fact that as the wave propagates, its energy converts into Joule heat that is generated by the conduction currents excited by the wave field.

First of all, let's illustrate that wave propagation through a homogeneous conductor is not related to the appearance of free electric charges. If we assume that there are no external electromotive forces in the conductor, we obtain:

$$\frac{\partial p}{\partial t} = -\text{div}\,\mathrm{j} = -\lambda\,\text{div}\,\mathrm{E} = -\frac{\lambda}{\varepsilon} = -\frac{4\pi\lambda}{\varepsilon}p \qquad (4.63)$$

The solution of this differential equation is

$$p = p_0\, exp\left(-\frac{4\lambda}{\varepsilon}t\right) \qquad (4.64)$$

where p_0 is an arbitrary constant.

Thus, even if free spatial charges are introduced into the conductor, the density of such charges will diminish to zero in the course of time in accordance with an exponential law. The higher the conductivity λ, the quicker the charges are dispersed. The electromagnetic field cannot create free spatial charges in the conductor at all, because if p = 0 at t_0, then, according to (4.64), it will stay at zero at all times.

Let's consider the monochromatic wave with a frequency of ω in metal; that is, let's assume

$$E = E_0(x,y,z)e^{i\omega t},\quad H = H_0(x,y,z)e^{i\omega t} \tag{4.65}$$

By introducing these expressions into Maxwell's equations, and, by assuming that according to (4.64) $p = 0$, we obtain after reducing by $e^{i\omega t}$

$$\left(\frac{i\omega t}{c}+\frac{4\pi\lambda}{c}\right)E_0 = rotH_0,\quad \frac{i\omega\mu}{c}H_0 = -rot\,E_0$$

$$div\,H_0 = div\,E_0 = 0. \tag{4.67}$$

These equations differ from the respective equations for dielectrics as follows: the multiplier iωt/c is replaced with iωt/s+4πλ/s in the first instance. In other words, these equations will coincide with the equations for waves in dielectrics if ε is replaced with ε' in the latter ones:

$$\varepsilon' = \varepsilon - \frac{4\pi\lambda}{\omega}i \tag{4.68}$$

Therefore, with regard to the propagation of the monochromatic waves, the conductor is an equivalent to the dielectric with complex dielectric permittivity ε'. Thus, considering waves in metal, we can use the results of the waves in dielectrics, only replacing ε with ε' in the formulas.

For example, the wave (complex) number k is determined according to the (4.11) formula

$$k'^2 = \frac{\varepsilon'\mu\omega^2}{c^2} = \frac{\varepsilon\mu\omega^2}{c^2} - \frac{4\pi\lambda\mu\omega}{c^2} \tag{4.69}$$

It is expedient to split κ' into real and imaginary parts:

$$k' = k - is, \ \ k^2 - S^2 = \frac{\varepsilon\mu\omega^2}{c^2}$$

$$2ks = \frac{4\pi\lambda\omega}{c^2} \tag{4.70}$$

Let's agree to take for *k* and *s* the positive roots of these equations. According to (4.70) and (4.14), the field of the plane monochromatic wave propagating along the *z* axis in the conductor will be represented by formulas

$$E = E_0 e^{i(\omega t - k'z)} = E_0 e^{-sz} e^{i(\omega t - kz)}$$

$$H = H_0 e^{-sz} e^{i(\omega t - kz)} \tag{4.72}$$

Thus, the complex character of the wave number *k'* means that absorption is present: the amplitude of the wave decreases exponentially as the wave propagates. At A (amplitude) = 0 the imaginary part *s* of the wave number *k'* becomes zero and the damping of waves stops.

According to (4.20), vectors E and H in the conductor are mutually perpendicular. Together with the wave direction, they form a right-handed system. However, vectors E and H have different phases in the conductor as opposed to similar ones in the dielectric. Indeed, by substituting ε with ε' in formula (4.20), we obtain

$$H = \sqrt{\frac{\varepsilon'}{\mu}}[nE] \tag{4.73}$$

As the multiplier ε'/μ is complex, the phase of vector H differs from phase E.

Displacement currents in metals are small if compared with the conduction of currents up to the frequencies that correspond to the infrared spectrum ($\nu = \omega/2\pi \sim 10^{14}\ s^{-1}$). Thus, the results reported here are only slightly different from the results reported above for lower frequencies.

Indeed, by solving equation (4.70) for k and s, we obtain

$$k^2 = \frac{\varepsilon\mu\omega^2}{2c^2}\left[\sqrt{1+\left(\frac{4\pi\lambda}{\varepsilon\mu}\right)^2}+1\right]$$

$$s^2 = \frac{\varepsilon\mu\omega^2}{2c^2}\left[\sqrt{1+\left(\frac{4\pi\lambda}{\varepsilon\omega}\right)^2}-1\right] \tag{4.74}$$

As was mentioned earlier, in metals $\varepsilon v << \lambda$; that is, $\varepsilon\omega < 2\pi\lambda$ up to $v \sim 10^{14}$ s^{-1}. Therefore, it is possible to neglect the unity in (4.74) if compared with $4\pi\lambda/\varepsilon\omega$ even for the light waves. Thus, with a reasonable degree of accuracy

$$k^2 = s^2 = \frac{2\pi\mu\lambda\omega}{c^2} \tag{4.75}$$

As was mentioned earlier, the depth of wave penetration into the metal is determined by

$$\delta = \frac{1}{p} = \frac{1}{s} \tag{4.76}$$

because the wave amplitude dies out at such a depth by ε times if compared with the amplitude at the surface. As according to (4.17) the wavelength, which we denote as l this time, is equal to $2\pi/k$ and as $\kappa = s$, then

$$\delta = \frac{l}{k} = \frac{l}{2\pi} \tag{4.77}$$

Thus, only 1/6 part of the wavelength fits into the specified distance – that is, *there is no spatial periodicity of the wave in the metal.*

The following table illustrates the depth of penetration into copper of fields with various frequency values for ω. In the table, l_0 means the length of respective wave in a vacuum: $l_0 = 2\pi c/\omega$.[12]

Table 4.2 Depth of fields' penetration into copper

l_0 [m]	10^{-6}	10^{-2}	10^{2}
δ [m]	$3.8 \cdot 10^{-9}$	$3.8 \cdot 10^{-7}$	$3.8 \cdot 10^{-5}$

The metal surface reflection of light is much more complex than reflection at the boundary of the dielectrics. Thus, for example, a linearly polarized wave becomes an elliptically polarized wave after being reflected from the metal (if the incidence angle differs from 90°). We will limit ourselves to the review of the simplest case: the normal incidence of the plane monochromatic wave travelling from a vacuum to the metal surface.

According to (4.50), the metal refraction index n' with regard to a vacuum will be

$$n' = \sqrt{\varepsilon'} \tag{4.79}$$

That is, it will have a complex value. For the normal incidence of the wave onto the metal, the amplitudes of the electric vector of reflected and refracted waves will be determined by formula (4.57):

$$E_0^r = \frac{1-n'}{1+n'} E_0, \quad E_0^g = \frac{2}{1+n'} \tag{4.79}$$

By assuming in the following formulas:

$$E_x = EE_0 e^{i(\omega t - k_0 z)}, \quad E_x^r \frac{1-n}{1+n'} E_0 e^{i(\omega t + k_0 z)},$$

$$E_x^g = \frac{2}{1+n'} E_0 e^{-sz + i(\omega t - kz)}. \tag{4.80}$$

The real part of these complex expressions is equal to

$$E_x = E_0 \cos(\omega t - k_0 z),$$

$$E_x^g = \left|\frac{2}{1+n'}\right| E_0 e^{-sz} \cos(\omega t - kz - \psi),$$

$$E_x^r = \left|\frac{1-n'}{1+n'}\right| E_0 \cos(\omega t + k_0 z - \varphi) \tag{4.81}$$

where the angles ψ and φ shall be determined from the relations:

$$\frac{1-n'}{1+n'} = \left|\frac{1-n'}{1+n'}\right| e^{-i\varphi}, \quad \frac{1}{1+n'} = \frac{1}{|1+n'|} e^{-i\psi} \tag{4.82}$$

where, for example, $|1 + n'|\cdot j$ is the modulus if a complex value 1 + n'. Thus, due to the complex character of *n'*, the phases of reflected and refracted waves will not, as with dielectrics, coincide at the interface with the phase of an incident wave. They will be displaced by angles φ and ψ respectively.

The average (throughout the flux density period) values of the energy in the incident and reflected waves, and reflection factor *r* will be equal to

$$\overline{S} = \frac{c}{8\pi} E_0^2, \quad \overline{S^r} = \frac{c}{8\pi} \left|\frac{1-n'}{1+n'}\right|^2 E_0^2$$

$$r = \left|\frac{1-n'}{1+n'}\right|^2 \tag{4.83}$$

As A for metals with powers of ten 10^{16} — 10^{17} in absolute units (s^{-1}), then $4\pi\lambda/W>1$ up to visible light frequencies. It follows that according to (4.68) and (4.78), the modulus of *n'* is much greater than unity.

Therefore, the reflection factor of metal surfaces *r* is close to unity. Thus, for example, even for the yellow spectrum of sodium, *r* is equal to 0.95 for Ag, 0.85 for Au, 0.83 for Al, 0.74 for *Cu*, etc.[13]

5

A New Theory of the Structure of Matter

5.1 Advancing the new theory

Let's first review the well-known facts. Figure 5.1 shows the structure of the atom in accordance with Niels Bohr's theory.[14]

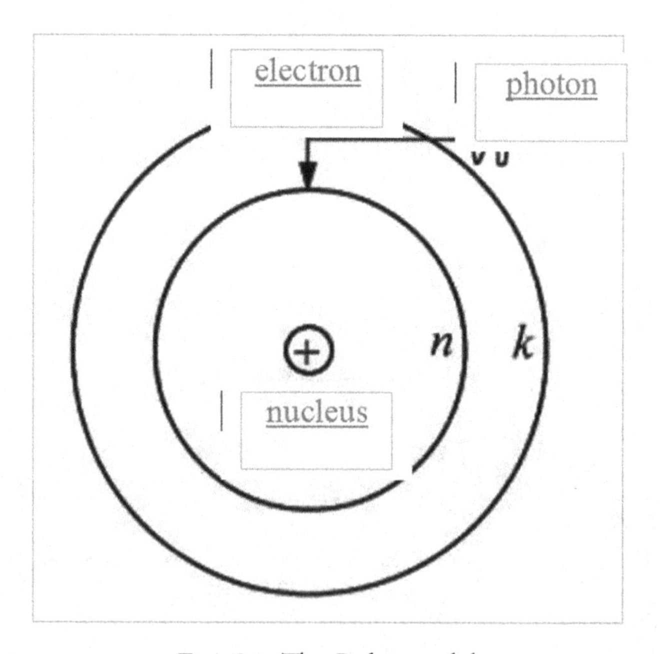

Fig. 5.1. The Bohr model.

The Bohr model of the atom is a semi-classical model of the atom suggested by Niels Bohr in 1913. He based his theory upon the atomic planetary model devised by Ernest Rutherford. However, according to classical electrodynamics, the electron of the Rutherford model should emit continuously while orbiting the nucleus. Therefore, it should lose energy and collapse into the nucleus very quickly. To overcome this difficulty, Bohr made an assumption that the electrons could only have certain stationary orbits. On these orbits, the electrons do not emit energy. Absorption or emission occur only when electrons jump from one orbit into another. The orbits can be considered stationary only if the electronic angular momentum equals the integer number of Planck's constant:

$$m_e v_r = n\hbar \tag{5.1}$$

Using this assumption and applying the laws of classical mechanics – namely, that the gravitational force of the electron towards the nucleus is equal to the centrifugal force influencing the orbiting electron – Bohr obtained the following values for the radius of the stationary orbit R_n and energy E_n of the orbiting electron:

$$R_n = \frac{4\pi \varepsilon_0 h^2}{Z m_e e^2}; \quad E_n = \frac{Z^2 m_e e^4}{32\pi^2 \varepsilon_0^2 h^2} \cdot \frac{1}{n^2} \tag{5.2}$$

Where m_e is the mass of an electron, Z is the number of protons in the nucleus, ε_0 is the dielectric constant, and e is the electron charge.

Exactly this expression can be obtained for the energy of the orbiting electron applying the Schrödinger equation, while at the same time solving the problem of electronic motion in the central Coulomb field.

The radius of the first orbit in the hydrogen atom $R_0=5.2917720859(36)\cdot 10^{-11}$ [m] is now called *the Bohr radius*, or the atomic unit of length.[15] It is widely used in the modern physics. The energy of the first orbit $E_0= 13.2$ eV is the ionization energy of the hydrogen atom.

It would be logical to suppose that there is no empty space in the atom, but there is a finer level of particles between the electrons, the proton, and the neutron. This level predetermines the occurrence of the wave disturbances for the larger particles. Upon receipt of an energy pulse, the electron jumps to the orbit which is closer to the nucleus. Then the respective photon influences the photon level which surrounds the atom. Moreover, if it is assumed that the photon is a component of the electron, it should be smaller and, therefore, the density should be higher.

Due to the wave motions in the atom, the electrons will have condensations and rarefactions, while it is most likely that the atom will have an integral multiplicity of frequencies. Such a multiplicity implies a sphere within a sphere.

The repeating pattern of travelling waves on the atom surface can be likened to the ocean waves on the surface of the planet. In its simplified form it can be presented as follows:

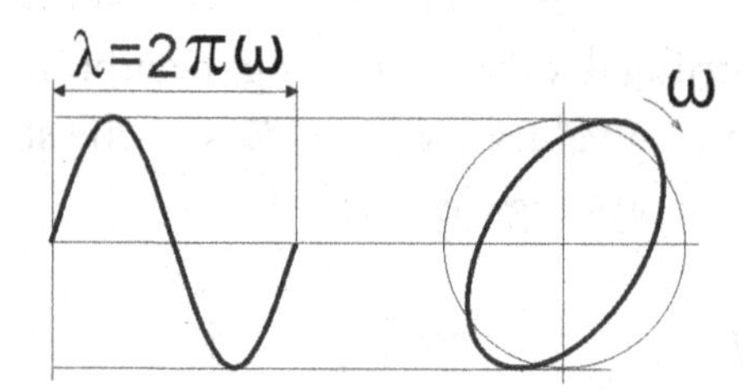

Fig. 5.2. The wave propagation effects in an atom.

We believe that the wavelength depends upon the oscillations of the particles at the other, finer level, but not of the particles at the given level. This explains logically the corpuscular theory (at the given level) together with wave theory (at the finer level). If at the given level it is a point, then at the finer level this point already has dimensions.

The system based upon three coordinates can be applied for only one level. Systems with 6, 12, or more coordinates can be applied for more purposes.

An excess mass occurs when the sum of the masses of the proton and the neutron is less than the mass of these protons and neutrons separately. It is connected with the binding energy at another level. An electron tube should provide a convincing example for the existence of ether in general and the photon level in particular. The tube is said to be empty inside; however, the electric current is somehow transferred from a cathode to an anode. It is also appropriate to mention the boundary, which is an area of space where the density of the medium or the space itself changes rapidly. The boundary itself does not have a physical thickness, but helps to explain the essence of the wave-corpuscle duality: that is, the medium, the wave, the boundary, and the body are all present at the same time.

The operation of a magnetohydrodynamic (MHD) generator indirectly supports energy conversion from a chemical level (molecular) into a radiation level (photon).

The idea that an electric current is a wave propagation through the photon field also makes a lot of sense. Let's suppose that two electrons are taken from the outer level of a metal – for example, copper, which conducts an electric current. This should lead naturally to the change of the material itself. It is generally accepted as true that only electrons transfer the electric current in metals. However, this position is challenged by the hypothesis advanced here.

In advancing a new explanation for the structure of matter, it is necessary, first of all, to provide the following data, which is referenced in:[16]

$m_e = 9.1 \cdot 10^{-31}$ *[kg] – the mass of the electron*

$m_p = 1836.1 \cdot m_e$ *– the mass of the proton*

$r_p = 6.7 \cdot 10^{-16}$ *[m] – the radius of the proton*

$h = 6.6 \cdot 10^{-34}$ *[J] – the Planck's constant.*

Let's suppose that an electron has a complex structure, similar to the structure of an atom of an isotope of hydrogen, deuterium. It will have three elements: a "proton", a "neutron" containing a "proton" and an "electron", and an "electron" at the outer orbit. Beta decay proves the complex structure of a neutron.

The beta decay of a neutron is an unprompted change of a free neutron into a proton, with the emission of a β-particle (electron) and an electron antineutrino:

$$ {}_0n^1 \rightarrow {}_1p^1 + e^- + \bar{\nu} \qquad (5.3)$$

The spectrum of kinetic energy of the emitted electron ranges from 0 to 782.318 keV. The free neutron lifetime is 885.7 ± 0.8 s (which corresponds to the decay half-life 613.9 ± 0.6 s). Precision measurements of the neutron beta-decay parameters (lifetime, angular correlation between particle momentum, and neutron spin) have an importance for determining the attributes of the weak interaction.

Frédéric Joliot-Curie predicted the beta decay of the neutron in 1934. A. Snell, J. Robson, and P.E. Spivak independently observed the process in 1948–1950.[17]

The complex structure of the electron is proved in the following article.[18]

"A new type of neutron decay, namely the radiative beta decay, was observed experimentally. Such a discovery became possible due to the development of low-energy particles detectors.

Life for the majority of currently known elementary particles is bright and transient. Born together with fellow particles after the reaction of the collision between protons and electrons, they cover a microscopic distance before decaying into other particles. Post-decay finite states (or, as physicists call them, the decay channels) might be different as long as the fundamental laws of physics (charge conservation law, energy conservation law, et cetera) are observed. Over a hundred decay channels are known for some particles.

Only an insignificant number of particles lives long enough to come in direct contact with the outside world. During their lifetime they manage to fly for a considerable distance: centimetres, metres, and, in very rare cases, kilometres. But their decay is very quick by human reckoning – in a split second.

The neutron is in a league of its own amongst this diversity. Firstly, it holds the record for a life-time amongst all elementary particles. On average it lives in a free state for 15 minutes, which is half a billion times more than its closest rival! Secondly, where its decay is concerned, observed results are amazingly elusive. One-and-only-one type of neutron decay has been observed so far, namely into proton, electron, and antineutrinio (see the review article of B.G. Yerozolimskiy. Neutron beta decay. UFN, 1975, vol. 116, issue. 1, p. 145-164).

After the study of this particle for over half a century, recently it looks like physicists have discovered a second type of neutron decay. The preprint nucl-ex/0512001, drafted by a research group consisting of Russian, Belgium, and German scientists, reports that the radiative beta decay of a neutron was successfully observed. In this instance the neutron decayed into a proton, an electron, an antineutrino, and a photon. The discovery of this rare mode of neutron decay was achieved through the identification of events involving triple coincidences: the simultaneous registration of electron and photon; and the measurement of the recoil momentum of the proton.

Generally speaking, this discovery is not surprising for theoreticians. It is known that photons can "exit" along with other particles in all types if reactions involving charged particles (protons and electrons are electrically charged). However, observation in the case of the neutron turned out to be technically very challenging. All exiting particles possess low levels of energy, and it is difficult therefore to catch trace of them with detectors.

In 2002 the same group failed to discover this mode of decay. Their recording equipment was not accurate enough to detect it. However, after the detectors were upgraded and data processing procedures were enhanced, the researchers found that free neutrons decay with the emission of a photon in one case out of three hundred on average.

The accuracy of this experiment is fragile. It might turn out (although it is unlikely) that the recorded "signal" is just a result of the random overlapping of background processes. However, the researchers note that it is possible to further improve the method, which would ensure 10% accuracy while measuring the probability of this decay."[19]

Thus, on the basis of experimental data and working logically, it is reasonable to assume that an electron is a complex formation consisting of photons. Let's try to hypothesise how many photons constitute an electron. It is obvious that the minimum number of elements required for a rigid structure is 3.

It follows that an electron can have one "neutron" of the second level (n_2), one "proton" of the second level (p_2) and one "electron" of the second level ($e-_2$). Therefore, an "electron" of the first level might contain 6 photons. A "proton" of the first level might contain 12 photons (2 "electrons"). A "neutron" might contain 18 photons ("proton" and "electron"). Thus, an atom of deuterium might have 6 +12 +18 = 36 photons in total (see Fig. 5.3).

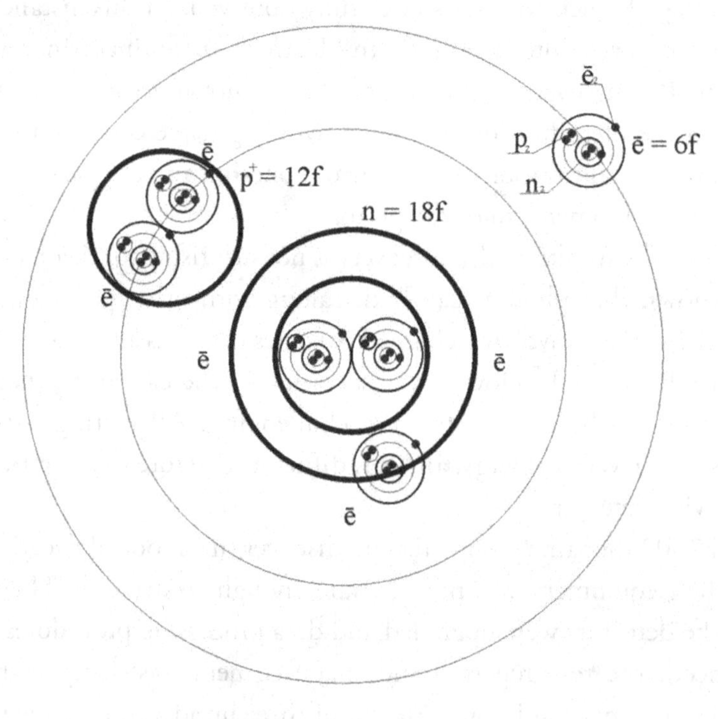

$$D = p^+ + n + \bar{e} = 18f + 12f + 6f = 36f$$

Fig. 5.3. The structure of an atom of deuterium according to the proposed data.

Earlier we provided information about the relation of the masses of electrons and protons ($m_p = 1836.1 \cdot m_e$). This seems to contradict our hypothesis, but if it is granted that the velocities of photons inside these particles are different, this contradiction can easily be accommodated.

Let's consider the structure of an atom in more detail, taking deuterium as an example. The electron, consisting of 6 photons, has a structure similar to that of an atom. However, the electron is the simplest element out of these three (neutron, proton, and electron) which comprise an atom.

When the conditions for chemical bonding between two electrons occur (during synthesis), then a proton is formed (see Fig. 5.4).

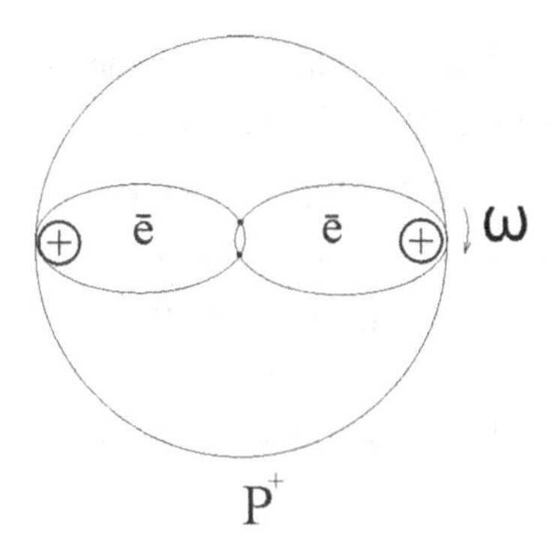

Fig. 5.4. The formation of a proton.

Figure 5.4 shows that common photons are located in the contact area of electrons. As a result, a negative charge is formed there, while a positive charge is formed on the external side of the chemically bonded electrons. The external sides of these electrons are the outer surface of the revolving proton. This means that the proton is positively charged for all interactions. Furthermore, the surfaces of the electrons that constitute the proton are of an ellipsoid shape due to the centrifugal force acting upon them.

Being positively charged, the proton is primed to attract a negatively charged electron. When this becomes possible, the proton creates a neutral energy particle – a neutron (see Fig. 5.5).

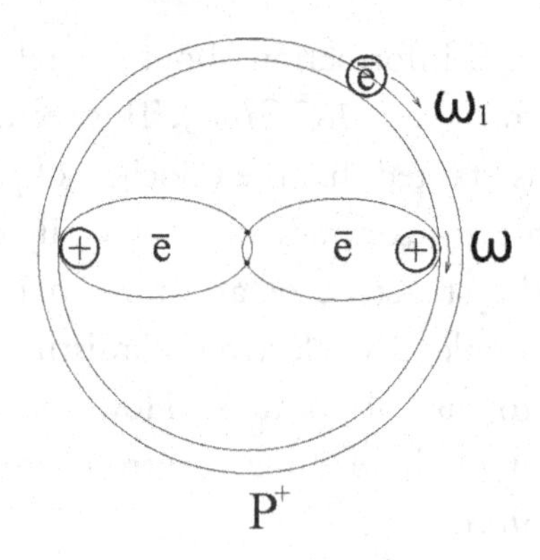

Fig. 5.5. The formation of a neutron.

Figure 5.5 shows that the electron revolves and slides along the boundary of the proton. The charge of the electron neutralizes the charge of the proton by spreading it over the surface of the neutron. In the continuing interactions, this electron easily escapes the neutron, which then reverts to being a proton again. The speed is $\omega_1 < \omega$.

This pattern of the structure is universal and repeats itself in both image and likeness at all times and at all levels:

$$\bar{e} \to \overset{+}{p} \to \overset{o}{n} \to A^{o}$$

$$\overset{-}{e_1} \to \overset{+}{p_1} \to \overset{o}{n_1} \to f$$

$$\overset{-}{e_2} \to \overset{+}{p_2} \to \overset{o}{n_2} \to \gamma f$$

It is well known that the radiation of an electromagnetic wave of a certain frequency is characterised by a given colour. However, the way such waves originate still calls for an explanation. After the analysis, we suggest the following hypothesis to explain the origin of electromagnetic waves that originate certain colours.

We have argued that the structure of the electron is similar to the structure of the deuterium atom: the photon in the electron (of the first level) is analogous to the electron (of the first level) in the atom; the proton of the second level in the electron (of the first level) is analogous to the proton (of the first level) in the atom; the neutron of the second level in the electron (of the first level) is analogous to the neutron (of the first level) in the atom.

Now, if we assume that the structure of the photon (the electron of the second level) is also analogous with the structure of the electron and with the structure of the deuterium atom, then the component particles of the photon will be a sum of the electron of the third level, the proton of the third level, and the neutron of the third level (see Fig. 5.8).

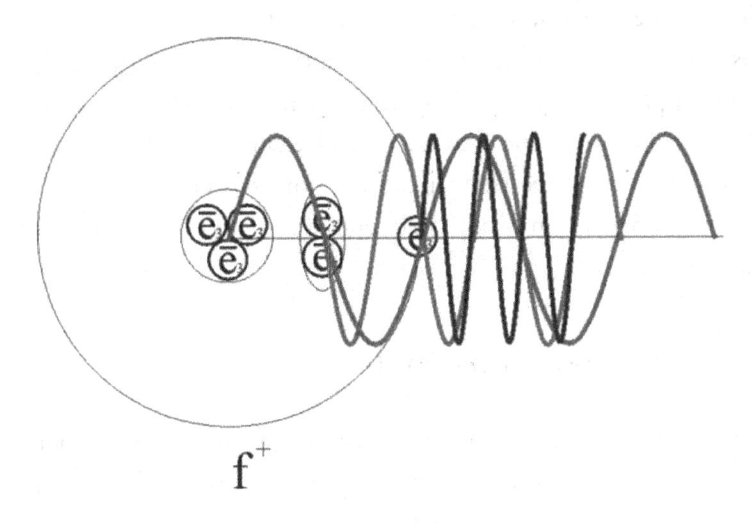

Fig. 5.8. The suggested structure of a photon.

Elements comprising the photon have different values of mass, hence, their various values of resonance oscillations. The mass of the third level neutron in the electron of the second level (photon) is minimal. Therefore, in all probability, the frequency of its oscillations will be minimal and will correspond to the colour red, for example. The proton of the third level, having middle values of mass and resonance frequency, will correspond to the colour green, say. And, finally, the third level electron, being the lightest, will probably correspond to the colour blue, say. The dominant background colour will depend upon the frequency of oscillations which prevails in the general set of all the frequencies. The overall background will reflect all the frequencies of the particles which comprise the photon. The various intensities of the oscillations of the respective frequencies will determine the colour pattern of a wave.

Note that the suggested colours (red, green, and blue) are purely conjectural and chosen for convenience. One could just as well apply any other choice – for example, brown, yellow, and purple. What is more important is the condition that photons transmit electromagnetic waves not at a single (photon) frequency, but as a set of frequencies of oscillation of electrons, protons, and the neutron of the third level. Only if this condition is fulfilled is it possible to transmit light energy with colour components where one wave of oscillations includes another. A further condition for the free transfer of electromagnetic waves is the multiplicity of all three frequencies to avoid the creation of obstacles. Figure 5.8 shows the multiple frequencies radiated by various particles of the photon.

5.2 The interaction of the levels

The interaction between the "material" level and the finer levels can be presented as follows. Material particles and bodies at a planet, such as the Earth, within a stellar system, such as the solar system, are surrounded by a photon level, which, in turn, is surrounded by a gamma-photon level, etc. Figure 5.9 represents this idea in a simplified form.

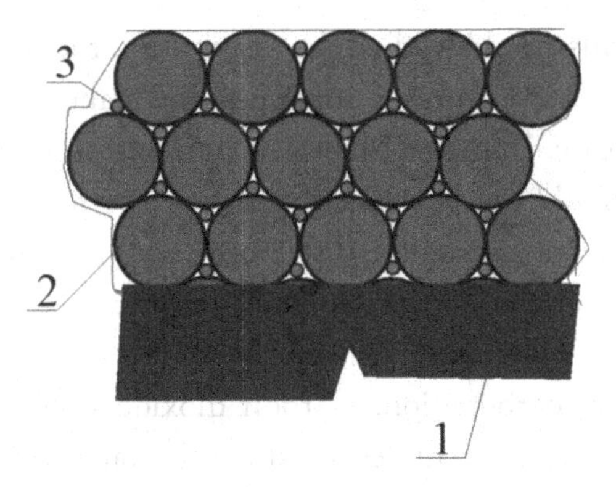

Fig. 5.9. The distribution of "non-material" and "material" levels.

Where:

1 – the given volume of a "material" body;

2 – the photons of a given volume at the photon level;

3 – the gamma-photons of a given volume at the gamma-photon level.

It is known that particles with a higher frequency of oscillation can transfer more energy. The gamma-photons are just such particles, as shown in Figure 5.9. They are denser because they serve as "construction material" for the coarser levels. Nevertheless, they do not have a dominant influence near the surface of a planet (such as the Earth) because they occupy no more than 10% (9.7%) of the surface area when compared with the photon level. This value is calculated as follows. The quantity of the volumes of the above determined photons is deduced from the given volume (see Fig. 5.9). The remainder will be the volume of the gamma-photon level in the given volume. The ratio of the remainder to the total volume of photons will be approximately 0.097 – that is, about 9.7%.

Having established that the most energy-intensive gamma-photon level does not play a potential key role during the interactions near the surfaces of planets, it is necessary to separately consider the issue of the levels of distribution. Of all the levels touched upon in this book (the number of levels may be greater in reality), the gamma-photon level seems more dense, as it is the finest level and, as mentioned earlier, serves as the "construction material" for the photon level.

Logically this would lead to the assumption that gamma-photons push out the photons from the surface of the planet. However, this is not quite the case. The coexistence of photons and gamma-photons in one space should not be viewed as an open system, where gaseous, liquid, and solid substances are distributed according to the following principle: dense at the bottom, loose at the top. It should be viewed as a closed, confined system. The entire process of coexistence can be likened to water carbonation. Carbon dioxide –the substance with lesser density – is injected under pressure into water (by analogy to our case, the gamma-photon level), which is contained in a confined space. Molecules of gas (in our case, the photon level) build in between the molecules of water within the confined space of a bottle (in our case, outer space). A compressor (pressurized gas cylinder, etc.) serves to create pressure for water carbonation. When the photon level is built into the structure of the gamma-photon level, this function is realized by a star, such as the Sun, which produces photons from gamma-photons. The stellar system is closed because it is confined from all sides by the pressure of the gamma-photon level.

At this point a question naturally occurs: how can this pressure be present if we don't feel it? The answer might be as follows. We don't feel this pressure much in the same way that a plaice does not feel the pressure of the huge column of water above it. Note also that sea water has, like the gamma-photon level, a density which is higher than that of the fish.

As we know, energy is transferred through the cosmos by means of waves. The speed of the wave propagation through a medium is its *refraction index*.

Let's extend our considerations in accordance with the diagram in Figure 5.10.

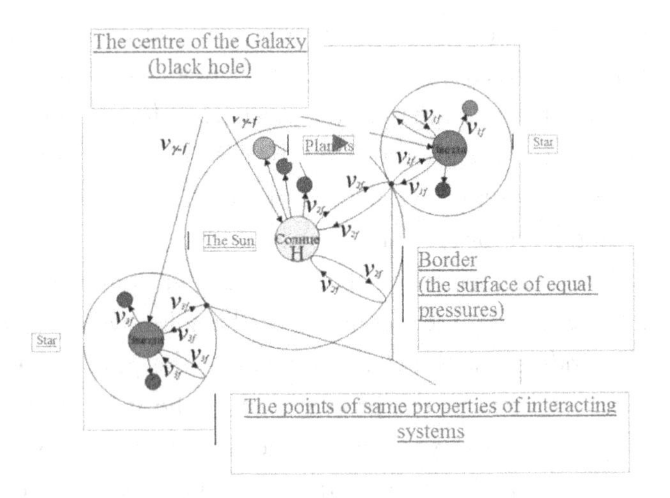

Fig. 5.10. Energy transfer in the cosmos
using the Galaxy as an example.

The centre of each stellar system is a star (the Sun, for the solar system). The star receives energy from the black hole through the gamma-photon level. Only stars can receive energy from the black hole because they act like a fine mesh sieve and can retain the gamma-photon radiation from the black hole. Planets act like coarse mesh sieves and, therefore, cannot retain the energy coming from the black hole. Another reason for this was mentioned earlier: the gamma-photon level does not play a decisive role for planets. Having received the energy from the black hole, the star undergoes transformations (for example, there are reactions in the Sun which result in high temperatures and hydrogen formation), and then synthesises photons from gamma-photons. The star then radiates energy at the photon level. The planets of a given stellar system (such as the solar system) then retain this energy and synthesise matter.

Although the gamma-photon does not play a key role for a planet, it does for a star such as the Sun. Unlike the planet, the star has no constant photon barrier which blocks the influence of the gamma-photon level, because the star intermittently emits photons to the areas with minimum potential pressure and extensive wave flows.

Furthermore, the frequency of oscillations of the emitted waves of the photon level is much less than the frequency of oscillations of the incoming waves of the gamma-photon level. Therefore, in between the emissions of waves of the photon level, the star receives a huge number of waves of the energy-intensive gamma-photon level. And this becomes the determining level for the star.

It should also be mentioned that every stellar system has its own frequency. For example, the frequency of the solar system is v_{2f}. The frequency depends upon the characteristics of the photons produced by the star and it leads to the propagation of the wave with a given resonance frequency within the boundaries of this stellar system only. This peculiarity is related to the occurrence of the maximum refraction index of the resonance wave at the system boundary. As a result, the wave returns to the star, its source – such as the Sun.

Having returned to the star, the wave penetrates into it until the maximum refraction index is achieved. But the wave cannot leave the star now, and the energy of the wave remains with the star, thus increasing its total energy.

Such a process is also possible for planets, which explains to some extent why certain layers of the surface of planets (including the Earth) get heated.

If we take the Earth as an example, this process can be described as follows. An electromagnetic wave arrives at a planet and penetrates into it until the refraction index reaches its highest value. The propagation of the wave ceases. The energy dissipates within the planet at the depth where the wave stopped its propagation, which leads to the "local" heating of the planet.

A boundary – that is, a surface of equal pressures – can be present at the areas of possible points of contact of various stellar systems. The boundaries of independent stellar systems have shapes that resemble spheres. Therefore, in the areas of possible contact, the properties of interacting systems might be the same. Such areas are a relatively small zone where the transition from one stellar system into another is possible without considerable energy loss.

Finalizing the review of the material at the level of macro-systems, it is well to bear in mind that a black hole is not "altruistic" at all. It absorbs the energy of radiation which has less frequency, and, in the course of time, it absorbs material space objects which earlier were created from its energy. With regular frequency it "shakes off" the space objects and converts them into energy. This is a closed process, shown in Figure 5.11 in a simplified form.

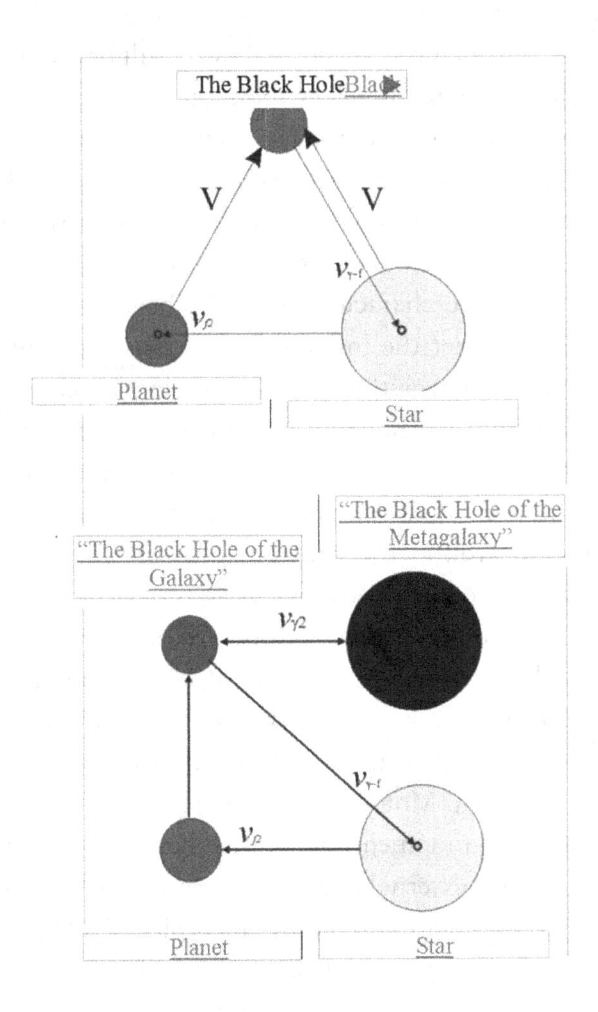

Fig. 5.11. The energy cycle in the Galaxy and Metagalaxy

5.3 A hypothesis for the formation of matter

It was mentioned earlier that stars, including the Sun, produce and emit photons. At the surface of the planet (including the Earth), the photons find themselves in an area of high potential pressure with an emerging resonance phenomena of photon and gamma-photon levels, for the most part. Such photons can be synthesised into formations which consist of a different number of original components.

Thus, material formations are created as a result of the synthesis of the photons. The question now arises about the mass and gravitational interaction of such formations, because they are "material" particles: electrons, protons, and neutrons. Wave interactions of various frequencies and directions are common to the "non-material" particles, such as photons and gamma-photons. As far as the "material" particles are concerned, they are characterized at the beginning by chaotic, disordered motion under the influence of various wave interactions of the finer levels. Later, with the accumulation of kinetic energy, such motion develops into the rotation of the formations with smaller mass around the formations with larger mass. The atomic nucleus is formed in this way after the proton accumulates enough energy for stable rotation around the neutron. The neutron also vibrates after getting energy from the wave interactions of the gamma-photon level. The radius of rotation of the proton around the neutron is not an exact circle, but it has a certain eccentricity value. This creates opportunities for the engagement of other particles with smaller mass, such as electrons, into the rotation. The rotations do not necessarily have the same direction. Moreover, the electrons have their own axes of rotation and angular momentum; that is, they spin. Of course, this occurs only if enough external energy is received. Rotating electrons form their orbits, which are limited by the orbits of rotating protons from the inside and by potential pressure of photon and gamma-photon levels from the outside. The number of electrons is distributed along the orbit according to its size, which therefore leads to the stability of the orbit. For example, only two electrons can consistently rotate at the first electronic energy level with its smaller orbit. As the size of further orbits increases, the number of electrons at the energy levels rises up to eight.

The number of rotating electrons differs according to the size, charge, and eccentricity value of the atomic nucleus. This also influences the number of electronic energy levels. Note that the intermediary orbits of the rotating electrons position themselves in such a way that the larger orbit encompasses the smaller one, because electrons on adjacent orbits repel each other. Electrons also repel each other within an orbit getting rotary motion because of the received energy. The outer orbit is a boundary one, so it divides various types of energy. Thus, everything inside the outer orbit is determined by the overall kinetic energy of motion of the electrons and by the atomic nucleus. Everything outside the outer orbit is determined by the overall potential and wave energy of the gamma-photon and photon levels. Furthermore, while potential energy of the photon and gamma-photon levels influences only the outer orbit, wave energy of the photon and gamma-photon levels penetrates inside the atom and leads to the increase of its overall kinetic energy.

Therefore, the atoms can be broadly viewed as energy vortices in an overall condensed medium. In other words, the atoms can be likened to accumulators of energy. The outer orbits are surfaces of equal values of the overall kinetic energy of the atom and the overall energy of the potential pressures of the photon and gamma-photon levels. Figure 5.12 provides a simplified representation of this description.

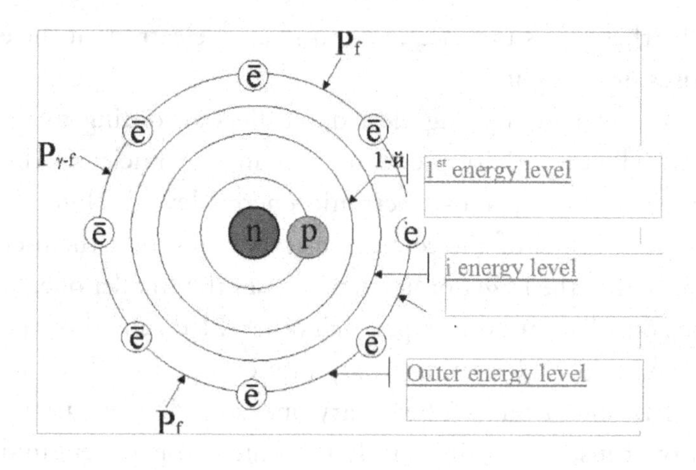

Fig. 5.12. The model of atom structure.

Where:

n – neutron;

p – proton;

$\bar{e}$ – electron;

P_f – the potential pressure of the photon level;

$P_{\gamma\text{-}f}$ – the potential pressure of the gamma-photon level.

However, the role of the outer orbits is not confined to the differentiation of various energy types but most likely to the accumulation of energy. With the inner orbits of the atom, an electron can jump one energy level down while another jumps one energy level up. The situation is different with the outer orbits. An electron jumps from the outer orbit to the adjacent lower orbit in the same way as happens for the inner orbits. However, it is extremely difficult for an electron to go above the boundary of the outer orbit. Thus, if the electron cannot leave the outer orbit, the energy should be accumulated inside the atom. However, conditions can be created for the electron to leave the outer orbit of the atom, which can happen discretely as follows. Having left the outer orbit, the electron loses energy outside the atom and immediately returns to its former place. As a result, electrons of the outer orbit can only "pulsate" around this orbit, which leads to further energy transfer to the photon and gamma-photon levels. The exception will be that of thermal emission.

5.4 The theoretical justification of the proposed theory

5.4.1 The effect of the electromagnetic field frequency on the dielectric permittivity of the dielectric

According to equation (5.4) the speed of the electromagnetic wave propagation depends upon the dielectric permittivity (ε) of the dielectric:

$$v = \frac{c}{\sqrt{\varepsilon}} \; [m \cdot s^{-1}] \qquad (5.4)$$

Where:
v – the speed of the electromagnetic wave propagation $[m \cdot s^{-1}]$,
c – the speed of light in vacuum $[m \cdot s^{-1}]$,
ε – dielectric permittivity.

The relative dielectric permittivity ranges from 2 to 8 (for a static field) for most solid or liquid dielectrics.[20] The value of the dielectric constant of water in a static field is quite high – approximately 80. Such values are high for substances with molecules having a large electric dipole. The relative dielectric permittivity of ferroelectric materials amounts to tens and hundreds of thousands.

It should be noted that the dielectric permittivity depends to a large extent upon the frequency of the electromagnetic field. This fact should always be taken into account, as reference tables usually provide data for a static field or low frequencies up to certain kHz without doing so. There are also optical methods for obtaining the relative permittivity based on a refraction index by means of ellipsometers and refractometers. The values obtained by the optical method (frequency 10^{14} Hz) will differ markedly from the data in the reference tables.

Let's take water as an example. For the static field (frequency equals zero), the relative dielectric permittivity under normal conditions will be approximately 80. This is true up to IR frequencies. Starting from around 2 GHz, ε_r begins to fall. For the optical band ε_r is around 1.8. This quite matches the fact that the value of the water refraction index is equal to 1.33 in the optical band.[21]In the narrow band, known as *the optical band*, the dielectric permittivity drops to zero. Basically, this is why a person can see in the Earth's atmosphere, which is saturated by water vapour. The properties of the medium change again as the frequency rises.

Thus, the speed of the electromagnetic wave propagation is not constant and depends upon frequency.

A proton consisting of two electrons is best suited for clear illustration of these features and provides for convenient calculations. Therefore, using the expression for relativistic mass, let's calculate the speed of the electromagnetic wave propagation for such a proton:

$$m_p = \frac{2m_e}{\sqrt{1-\frac{v^2}{c^2}}} \quad (5.5)$$

After transformation we obtain:

$$v_e = \sqrt{c^2\left(1-\left[\frac{2m_e}{m_p}\right]^2\right)} \quad (5.5)$$

Having applied that $m_p/m_e = 1836$, we obtain:

$$v_e = c\cdot 0.99986 = 29979245\cdot 0.99986 = 29975163.3\left[m\cdot s^{-1}\right]$$

Let's now calculate the frequency of the electromagnetic wave propagation (oscillation frequency of an electron) in a proton:

$$v_p = v_e / 2\pi r_p,\ [s^{-1}] \quad (5.6)$$

where:

v_e – the speed of the electromagnetic wave propagation in a proton [m·s^{-1}]

r_p – the radius of a proton [m]

$v_p = 2.9975163 \cdot 10^8 / 2\pi \cdot 6.7 \cdot 10^{-16} = 2.9975163 \cdot 10^8 / 4.2076 \cdot 10^{-15} =$
$= 0.7122 \cdot 10^{23}$ [s^{-1}]

Then let's calculate the length of the electromagnetic wave in a proton λ [m]

$\lambda = c' / v_p$, (5.7)

given that $c' = v_e$

$\lambda_p = v_e / v_p = 2.9975163 \cdot 10^8 / 0.7122 \cdot 10^{23} = = 4.2088 \cdot 10^{-15}$ [m]

The values obtained coincide with the proton equator up to four decimal places. That is, the electromagnetic wave in a proton travels along its ball-like surface.

This fact testifies that the proposed model of the structure of elementary particles, as exemplified by the proton, does not contradict our calculations.

5.4.2 The study of the parameters of elementary particles

An electron and other particles are represented in modern physics by electromagnetic waves. This approach is dualistic and needs a radical rethink. If it is held that it is impossible to locate an electron in space and specify its size, then the result is a physical paradox. It is a given that the primary characteristic of a material object is its boundary with the surrounding medium. Considering the spin of a particle as its rotational impulse around one of three coordinate axes (the two remaining coordinate axes of an electron determine its charge), and taking into account the well-known physical value of the spin, the radius of the electron has been calculated as the projection of the central moment of the particle's rotation against one of its spin axes. It is also possible to calculate more precisely the values of the spins of an electron, a neutrino, and, hence, of other subatomic particles – that is, fermions.

Let's calculate the spin value using Planck's constant:

$$S_0 = ¼\ m_0 \cdot c\ r_e = h/(4\pi\sqrt{c_{dless}})\ [kg \cdot m^2 \cdot s^{-1}] \quad (5.8)$$

Where:

m_0 – the mass of a particle at rest [kg],

c – the speed of light [m·s^{-1}],

r_e – the radius of the electron [m],

h –Planck's constant = 6.6261 · 10^{-3} [J],

c_{dless} – the speed of light (dimensionless quality),

$S_0 = 6.6261 \cdot 10^{-34}/(4\ \pi\sqrt{2.998 \cdot 10^8}) = 3.04539 \cdot 10^{-39}$ [kg·m^2·s^{-1}].

After obtaining the spin value, it is possible to calculate the radius of the electron r_e:

$$r_e = 4 \cdot S_0/(m_e \cdot c)\ [m]$$

$$r_e = 4 \cdot 3.04539 \cdot 10^{-39}/(0.911 \cdot 10^{-30} \cdot 2.99792 \cdot 10^8) = 4.4605 \cdot 10^{-17}\ [m]$$

To understand interaction of an electron with a medium, it is important to get another assessment of the electron's size by calculating the ratio of its volume to its surface. This approach complies with the above law of the world's evolution: the principle of the maximum of configuration energy. A free electron creates spherical power fields (electrical and gravitational) around itself. Each field acts like a central field with regard to any external particle interacting with the electron. Hence, these fields are functions of the electron's volume. However, any interaction occurs through a surface. Therefore, the quotient of the electron volume by its surface is of particular importance for every external interaction. It is either the fraction of the volume of the electron's absolute property through its surface unit or the relative amount of force field radiated through the surface unit of the electron. The mass of the electron is a quadratic function of the angular speed, whereas the interaction of two particles is a quadratic function of the electric charge or the mass and distance of the particles. Therefore, we are interested in a squared ratio between the volume V_e and the surface S_e of the electron:

$$(V_e/S_e)^2 = (4/3\pi\, r_e^{\,3})/\{(4\pi\, r_e^{\,2})\}^2 \text{ [m]} \qquad (5.7)$$

where r_e is the radius of the electron [m].
It follows that:
$(V_e/S_e)^2 = r_e\ 2/9 = 2.210\cdot10^{-34}$ [m],
$3(V_e/S_e)^2 = 6.630\cdot\ 10^{-34}$;
$h = 6.626176\cdot10^{-34}$ [J·s],
where h is Planck's constant.
It follows that without consideration of the dimensions:

$$(V_e/S_e)^2 = h/3, \qquad (5.8),$$

$$r_e = (3h)^{1/2} = 4.4585\cdot10^{-17} \text{ [m]}.$$

The radius of the electron turns out to be directly related to Planck's constant, which means that they are physically related, and yet again this confirms the detection of the size of the electron. And this is not a coincidence. This is a dramatic confirmation of the singleness and relative simplicity of the nature's laws.

Using the calculations of the relation between wavelength, speed of light, and frequency, we obtain the following values:

$$\lambda = c/v \text{ [m]} \quad (5.9)$$

$$\lambda_{red} = 0.7\cdot10^{-6-}\text{[m]}\ \lambda_{green} = 0.5\cdot10^{-6}\text{ [m]}\ \lambda_{blue} = 0.3\cdot10^{6}\text{ [m]}$$

$$v = c/\lambda \text{ [Hz]} \qquad (5.10)$$

where λ is the wavelength of photon radiation, and v is the frequency of photon radiation.

$v_{red} = 4.3 \cdot 10^{14}$ [Hz]
$v_{green} = 6.0 \cdot 10^{14}$ [Hz]
$v_{blue} = 10^{15}$ [Hz]

The energy, E, of the photon is calculated using the formula:

$$E = h \cdot v \qquad 5.11)$$

where:
E – the energy of the photon,
h – Planck's constant,
v – the frequency of the photon radiation.

$E_{red} = 6.6 \cdot 10^{-34} \cdot 4.3 \cdot 10^{14} = 28.38 \cdot 10^{-20}$ [J]
$E_{green} = 6.6 \cdot 10^{-34} \cdot 6.0 \cdot 10^{14} = 39.6 \cdot 10^{-20}$ [J]
$E_{blue} = 6.6 \cdot 10^{-34} \cdot 10^{15} = 66.0 \cdot 10^{-20}$ [J]

On the other hand, the energy, E, of a moving particle is determined by its speed (v) and mass (m):

$$E = (m \cdot v^2)/2 \text{ [J]} \qquad (5.12)$$

The speed of the wave of photons is constant at this level and is equal to the speed of light. Therefore, the mass of the photons changes. It is possible to calculate the mass of the photons by equating the energy using formulas (5.11) and (5.12).

$$E = h \cdot v = (m \cdot v^2)/2 \Rightarrow m = 2 \cdot h \cdot v/v^2;$$

We obtain the following mass value:

$v = c = 3{\cdot}10^{-8}$ [m·s^{-1}]

$m_{red} = 2{\cdot}28.38{\cdot}10^{-20} / 9{\cdot}10^{16} = 6.3{\cdot}10^{-36}$[kg]
$m_{green} = 2{\cdot}39.6{\cdot}10^{-20} / 9{\cdot}10^{16} = 8.8{\cdot}10^{-36}$[kg]
$m_{blue} = 2{\cdot}66.0{\cdot}10^{-20} / 9{\cdot}10^{16} = 14.67{\cdot}10^{-36}$[kg]

The mass of the photon as a ball can be calculated using the formula below:

$$m_f = (4/3){\cdot}\pi{\cdot}r^3\rho_f \ [kg] \tag{5.13}$$

where ρ_f is the density of the photon [$kg{\cdot}m^{-3}$].

$$r = \sqrt[3]{\frac{3m_f}{4\pi\rho}} \tag{5.14}$$

We assumed that protons and neutrons consist of photons. Therefore, the densities of photons and neutrons are approximately the same.

By substituting $\rho_f = \rho_n = 2.8{\cdot}10^{17}$ [kg/m^{-3}], that is, the mean density of a neutron star, we obtain:

$$r_f = \sqrt[3]{\frac{3\cdot 6.3\cdot 10^{-36}}{4\pi\cdot 2.8\cdot 10^{17}}} = 0.81319482 \cdot 10^{-17} \text{ [m]}.$$

Calculation results:

$r_p = 6.7{\cdot}10^{-16}$ [m] – the radius of the proton

$r_e = 4.4585{\cdot}10^{-17}$ [m] – the radius of the electron

$r_f = 0.81319482 \cdot 10^{-17}$ [m] – the radius of the photon

5.5 Matter and its levels

It has been mentioned that, while developing the periodic table, D.I. Mendeleev made provisions for a zero group of elements which are lighter than hydrogen. This provision fully complies with the hypothesis about the multiple-level nature of existing matter. Table 5.1 summarizes this statement.

Table 5.1

Level	Frequency
Micro-levels	$\nu > \nu_{\gamma-f}$
Gamma-photon	$\nu_{\gamma-f}$
Photon	ν_f
Atomic	ν_a
Molecular	ν_m
Macro-levels	$\nu_M < \nu_m$

We believe that each level has been adequately described in this book. The table illustrates that each level generates specific oscillation frequencies. But each level can also transmit the oscillations of another level. It becomes clear that new formations with new oscillation frequencies originate in each level. It should also be noted that the size of formations increases with each level, whereas the oscillation frequency decreases with each level.

Each level is mainly influenced by adjacent levels. Therefore, the properties of a level are determined, among other things, by the properties of adjacent levels. The coarser level consists of the particles of the finer level. For example, a proton, being a complex structure similar to an atom, consists of the particles of the finer level. We have assumed that the particles of the finer level serve as a source for formations at a higher level, and that each object at any level consists of three particles.

It should be emphasised that the natural (resonance) oscillation frequency of a given level is specific for this given level only, and not for the finer or coarser levels.

The transfer of energy between levels is not the same: more energy is transferred from the finer to the coarser levels than in the opposite direction.

Referring to the energy transfer between the levels, it makes better sense to talk about such transfers being through impulses.

It should also be noted that each level is characterized by its own speed of wave propagation and its own Planck's constant.

As of today, it is impossible to define either the ultimate fine level or the ultimate coarse level.

Conclusion

Here, at the end of our book, we should like to refer back to Chapter 5. Using its findings we arrive at some very interesting conclusions. Based on equation (5.7) $\lambda = c' / \nu_p$ and the suggested structure of the electron, it is possible to calculate the frequency of rotation of photons in the electron:

$\nu_f = c / 2\pi r_f$ (P.1)

$\nu_f = 2.9975 \cdot 10^8 / 6.28 \cdot 0.8 \cdot 10^{-17} = 5.966 \cdot 10^{24}$ [s^{-1}]

Knowing the frequency of rotation of photons, which is the frequency of the interactions in the next gamma-photon level, it is possible to determine the speed of wave propagation through this level:

$c_\gamma = c \cdot (\nu_f / \nu_\gamma) = 2.9975 \cdot 10^8 \cdot (5.966 \cdot 10^{24} / 3 \cdot 10^{15}) = 6 \cdot 10^{17}$ [m/s]

where ν_γ is the frequency of gamma emission.

The result obtained, combined with Einstein's equation $E = m \cdot c^2$, enables an estimation of the value of the energy confined within the unit of mass at the gamma-photon level.

As the energy is proportional to the squared speed of wave propagation through a given level, it is possible, using the ratio of squared speed, to obtain the relation of energy to the unit mass in the adjacent levels.

$$E_\gamma / E = c_\gamma^2 / c^2 \cong 36 \cdot 10^{34} / 9 \cdot 10^{16} = 4 \cdot 10^{18}$$

Thus, the specific energy that can be obtained as a result of the electron decay exceeds by 10^{18} times the specific energy of the nuclear decay.

By analysing the equation of the neutron β–decay (5.3):

$$_0n^1 \rightarrow {}_1p^1 + \bar{e} + \bar{\upsilon} \qquad (5.3)$$

It is possible to hypothesise that the hydrogen atom H_1^1 has been created after partial synthesis of a proton and an electron.

If the electron in the hydrogen atom changes its orbital velocity – for example, due to compression after a nuclear explosion – it will be attracted by the proton, and the neutron will be synthesised:

$$H_1^1 \leftrightarrow n^0$$

Thus, the hydrogen H_1^1 is the simplest object at a molecular level and the most complex at the level of "elementary" particles:

$$\ldots \leftarrow H_1^1 \leftrightarrow n^0 \rightarrow p_1^1 + \mathrm{e} + \nu \ldots$$

The following not impressive conclusion can be made based on the proposed structure of the electron and the proton. The charge sign depends upon the direction of rotation of the object, and it is possible to calculate the mass according to the Lorentz equation (see Fig. C.1).

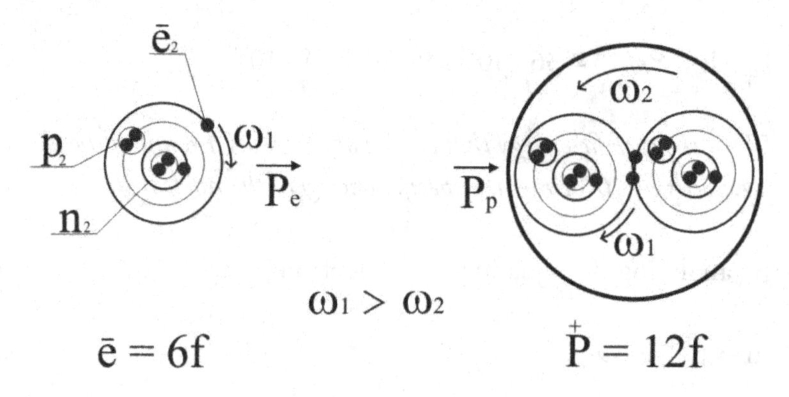

Fig. C.1. The structure and formation of the charge of the electron and the proton.

The mass of the proton is 1836 times greater than the mass of the electron. Hence, it is possible to calculate the velocity of two electrons rotating around the common centre of gravity in the proton:

$$m_p = \frac{2m_e}{\sqrt{1-\frac{v^2}{c^2}}} \tag{5.5}$$

After transformation we obtain:

$$v_e = \sqrt{c^2\left(1-\left[\frac{2m_e}{m_p}\right]^2\right)} \tag{5.5}$$

$$v_e = c \cdot 0.99986 = 29979245 \cdot 0.99986 = 29975163.3\left[m \cdot s^{-1}\right]$$

$v_p = v_e / 2\pi r_p$ [s^{-1}] (5.6)

$\nu_p = 2.9975163 \cdot 10^8 / 4.2076 \cdot 10^{-15} = 0.7122 \cdot 10^{23}$ [s^{-1}]
$\boldsymbol{\omega_1} = \nu_f = 5.966 \cdot 10^{24}$ [s^{-1}]
$\boldsymbol{\omega_2} = \nu_p = 0.7122 \cdot 10^{23}$ [s^{-1}]

Thus, an electron is a divergent spherical wave with a natural frequency $\omega_1 = 5.966 \cdot 10^{24}$ [s^{-1}], whereas a proton is a convergent spherical wave with frequency $\omega_2 = 0.7122 \cdot 10^{23}$ [s^{-1}].

Endnotes

1. Karpova, T.V., *Kontseptsii sovremennogo yestestvoznaniya. Konspekt lektsiy [Concepts of Modern Natural Science. A Summary of Lectures]* (In Russian.), p 5.
2. Einstein, A., *Sobraniye nauchnykh trudov v chetyrekh tomakh [A Collection of Scholarly Works in Four Volumes],* volume 2, (In Russian.), pp 167-183.
3. Einstein, A., *Sobraniye nauchnykh trudov v chetyrekh tomakh [A Collection of Scholarly Works in Four Volumes],* volume 2, (In Russian.), p 170.
4. Einstein, A., *Sobraniye nauchnykh trudov v chetyrekh tomakh [A Collection of Scholarly Works in Four Volumes],* volume 2, (In Russian.), p 192.
5. Einstein, A., *Sobraniye nauchnykh trudov v chetyrekh tomakh [A Collection of Scholarly Works in Four Volumes],* volume 1, (In Russian.), p 101.
6. Einstein, A., *Sobraniye nauchnykh trudov v chetyrekh tomakh [A Collection of Scholarly Works in Four Volumes],* volume 1, (In Russian.), p 124.
7. Karpova, T.V., *Kontseptsii sovremennogo yestestvoznaniya. Konspekt lektsiy [Concepts of Modern Natural Science. A Summary of Lectures]* (In Russian.), p 82.
8. Karpova, T.V., *Kontseptsii sovremennogo yestestvoznaniya. Konspekt lektsiy [Concepts of Modern Natural Science. A Summary of Lectures]*, p 8.

9. Lenta.ru *"Fiziki poluchili fotonnuyu materiyu" [Physics Create Photon Matter]*, http://lenta.ru/news/2013/09/26/boxofphotons/ (in Russian)
10. Yavorskiy, B.M. and Detlaf, A.A., *Spravochnik po fizike [The Physics Handbook]* (In Russian.), p 803.
11. Yavorskiy, B.M. and Detlaf, A.A., *Spravochnik po fizike [The Physics Handbook]* (In Russian.), p 571.
12. Yavorskiy, B.M. and Detlaf, A.A., *Spravochnik po fizike [The Physics Handbook]* (In Russian.), p 425.
13. Yavorskiy, B.M. and Detlaf, A.A., *Spravochnik po fizike [The Physics Handbook]* (In Russian.), p 538.
14. Matveyev, A.N., *Atomnaya fizika. Uchebnoye posobiye dlya studentov vuzov [Atomic. Physics. Tutorials for Students of Higher Education]* (In Russian.), p 48.
15. Matveyev, A.N., *Atomnaya fizika. Uchebnoye posobiye dlya studentov vuzov [Atomic. Physics. Tutorials for Students of Higher Education]* (In Russian.), p 23.
16. Sheshukov, V.V. and Groo V.Ya, *Novyy vzglyad na stroyeniye materii [A New Approach to the Structure of Matter]* (In Russian.), p 91.
17. Matveyev, A.N., *Atomnaya fizika. Uchebnoye posobiye dlya studentov vuzov [Atomic. Physics. Tutorials for Students of Higher Education]* (In Russian.), p 396.
18. Ivanov, I.S., *Neitroni raspodautsya i s izlucheniem na fotoni [Neutrons can also decay with photon emission]* (in Russian), *News of Science*, http://elementy.ru/news/165038, accessed 20 January 2013
19. Ivanov, I.S., *Neitroni raspodautsya i s izlucheniem na fotoni [Neutrons can also decay with photon emission]* (in Russian), *News of Science*, http://elementy.ru/news/165038, accessed 20 January 2013
20. Yavorskiy, B.M. and Detlaf, A.A., *Spravochnik po fizike [The Physics Handbook]* (In Russian.), p 346.
21. Yavorskiy, B.M. and Detlaf, A.A., *Spravochnik po fizike [The Physics Handbook]* (In Russian.), p 535.

References

1. Einstein, A., *Sobraniye nauchnykh trudov v chetyrekh tomakh [A Collection of Scholarly Works in Four Volumes]* (Moscow: Nauka, 1965 – 1967). (In Russian.)
2. Ivanov, I.S., *Neitroni raspodautsya i s izlucheniem na fotoni [Neutrons can also decay with photon emission]*(in Russian), *News of Science,* http://elementy.ru/news/165038
3. Karpova, T.V., *Kontseptsii sovremennogo yestestvoznaniya. Konspekt lektsiy [Concepts of Modern Natural Science. A Summary of Lectures]* (Moscow, AST, 2010). (In Russian.)
4. Lenta.ru *"Fiziki poluchili fotonnuyu materiyu" [Physics Create Photon Matter],* http://lenta.ru/news/2013/09/26/boxofphotons/ (in Russian)
5. Matveyev, A.N., *Atomnaya fizika. Uchebnoye posobiye dlya studentov vuzov [Atomic. Physics. Tutorials for Students of Higher Education]* (Moscow: Vysshaya shkola, 1989). (In Russian.)
6. Sheshukov, V.V. and Groo V.Ya, *Novyy vzglyad na stroyeniye materii [A New Approach to the Structure of Matter]* (Chelyabinsk: Dom Pechati, 2011). (In Russian.)
7. Yavorskiy, B.M. and Detlaf, A.A., *Spravochnik po fizike [The Physics Handbook]* (Moscow: Nauka, 1991). (In Russian.)

About the Author

Vladimir Groo was born in 1952 in a village near Chelyabinsk, Russia. In 1979 he graduated from The Ural State University of Economics. Thanks to outstanding teaching at the university, he formed a great interest in inorganic chemistry, biology, and physics.

In 2002, Groo published his first work, *The Paradigm of Creation.* He published "Quantum and the Structure of Photon Field" in 2010, and a year later wrote "A New Approach to the Structure of Matter".

Throughout these years, Groo analyzed and systemized the knowledge he'd formed from empirical research. Challenging life dilemmas and thoughts on the universe brought the author to Lutheran Church. In 1999 he became the chairman of his church council. Thorough analysis of the Bible allowed the author to look at the universe from a different perspective. This combination of life experience and empirical scientific research instilled in Groo the confidence to write a book building on the foundation of a general theory of the universe. Groo hopes that a theoretical understanding of the surrounding world structure will help to finally solve mankind's problem of limited energy sources.

About the Book

Fundamentals of the General Theory of the Universe builds upon knowledge empirically obtained by mankind. In doing so, it interprets phenomena that have until now lacked definitive explanation, such as superconductivity and propagation of electromagnetic waves in conductor and in vacuum; the book also resolves the dilemma of the well-known wave-practical dualism. For the first time ever, this book answers the question, "What is gravitational interaction?". The book points out the connection between our material world and the world of elementary particles, as exemplified by $n_0 \leftrightarrow {}^1H_1$. Here one will find classification of matter and its creation. The author underscores the universal character of the laws of energy distribution at all matter levels. The levels themselves are triune and built in "the image and after the likeliness". This book is the first attempt ever at combining all natural sciences and stepping out of the bounds of generally accepted concepts.